LEARNING FROM OTHERS

Science & Technology Education Library

VOLUME 8

SERIES EDITOR

Ken Tobin, *University of Pennsylvania, Philadelphia, USA*

EDITORIAL BOARD

Dale Baker, *Arizona State University, Tempe, USA*
Beverley Bell, *University of Waikato, Hamilton, New Zealand*
Reinders Duit, *University of Kiel, Germany*
Mariona Espinet, *Universitat Autonoma de Barcelona, Spain*
Barry Fraser, *Curtin University of Technology, Perth, Australia*
Olugbemiro Jegede, *The Open University, Hong Kong*
Reuven Lazarowitz, *Technion, Haifa, Israel*
Wolff-Michael Roth, *University of Victoria, Canada*
Tuan Hsiao-lin, *National Changhua University of Education, Taiwan*
Lilia Reyes Herrera, *Universidad Autónoma de Colombia, Bogota, Colombia*

SCOPE

The book series *Science & Technology Education Library* provides a publication forum for scholarship in science and technology education. It aims to publish innovative books which are at the forefront of the field. Monographs as well as collections of papers will be published.

Learning from Others

International Comparisons in Education

Edited by

DIANE SHORROCKS-TAYLOR

and

EDGAR W. JENKINS

University of Leeds,
U.K.

KLUWER ACADEMIC PUBLISHERS
DORDRECHT / BOSTON / LONDON

A C.I.P. Ctalogue record for this book is available from the Library of Congress.

ISBN 0-7923-6343-4

Published by Kluwer Academic Publishers,
P.O. Box 17, 3300 AA Dordrecht, The Netherlands.

Sold and distributed in North, Central and South America
by Kluwer Academic Publishers,
101 Philip Drive, Norwell, MA 02061, U.S.A.

In all other countries, sold and distributed
by Kluwer Academic Publishers,
P.O. Box 322, 3300 AH Dordrecht, The Netherlands.

Printed on acid-free paper

Printed in the Netherlands.

CONTENTS

Introduction

Diane Shorrocks-Taylor
School of Education, University of Leeds, UK

In September 1998, a conference was held at the University of Leeds entitled 'International comparisons of pupil performance: issues and policy'. It was arranged by two groups within the School of Education at the University, the newly formed Assessment and Evaluation Unit and the Centre for Studies in Science and Mathematics Education. The joint interest in international comparisons of performance had itself arisen from earlier involvement in a follow-up study of the 1995 TIMSS work in England, reported in a later chapter in this book, in which the TIMSS assessment outcomes were studied alongside the outcomes from the National Curriculum testing programme in England.

Some of the results of this investigation had proved both interesting and challenging so the decision was made to promote wider discussion of some key issues by inviting contributors from all over the world to a meeting the major aims of which were to promote an exploration of:

- the theoretical foundations of international comparative studies of student performance;
- the practical problems of carrying out such studies;
- the appropriateness of the assessment models and approaches used in international comparisons;
- the role of international comparative studies in raising standards of student performance;
- and how international studies affect the shaping of national policy on education.

Foremost, however, was the emphasis on evaluating the strengths and limitations of such comparisons and an attempt to investigate the effects of the outcomes on practitioners and policy-makers alike in participating countries.

Although the contents of this book are closely linked to the inputs to the 1998 conference, they do not constitute the conference proceedings. Instead,

D. Shorrocks-Taylor and E.W. Jenkins (eds.), Learning from Others, 5–11.

the editors took the decision to produce an edited volume of articles, mostly based upon conference presentations but not always so. Composing the book in this way allowed papers to be selected for their appropriateness to the theme and it also permitted some updating of material which, at the time of the conference, was necessarily provisional. The contributors have therefore been given the freedom to re-cast their contributions in the light of both conference discussion and further reflection or to produce something completely different, although in tune with the overall theme of the meeting.

The aim of Chapters 1 to 6 is to take further the examination of the theoretical underpinnings of international comparative studies of performance and explore some of the practical problems associated with them. Chapter 1 serves as an introduction to these two broad issues. Chapters 2 and 3 are concerned with the largest international comparative study to date, namely the 1995/9 TIMSS.

It is clear that later international surveys by the International Evaluation of Achievement have benefited from the mistakes and shortcomings of the earlier attempts. For instance, the Second International Mathematics Survey (SIMS) took much greater account of the influences of curricula, teaching approaches, school organisation, teacher training and student motivation than did the earlier study (FIMS). What this study also revealed, however, were the problems generated by the sheer scale of the enterprise: a wide range of data collected from many thousands of schools and students requires both efficient data management systems and a clear rationale for posing and answering key questions from the data. This study also had a longitudinal element to it, in that some countries assessed students at the beginning and end of the same school year, revealing some instability in the scoring which in itself raises most interesting assessment questions.

The surveys of science attainments in the early 1970s (the First International Science Study) and the early 1980s (the Second International Science Study) also showed evidence of learning the lessons of the earlier investigations and have increasingly sought to contextualise the outcomes in parallel ways to the mathematics studies.

In the TIMSS study, as Martin and Mullis outline in Chapter 2, many new countries came on board, not least because of generous funding from the USA and Canada, but this in itself introduced new complexities, particularly for the item banks and curriculum matching exercise. In particular, the novel aspects that were introduced included:
- more open-ended response questions alongside the traditional multiple-choice ones;
- the assessment of practical skills – performance assessment;
- measures of 'mathematical and scientific literacy' at the end of the school years;

– and videos of classroom practice in selected countries.

Chapter 2 also presents some of the findings, with special reference to the USA. These indicate that in relation to the achievement testing, the USA sample of fourth grade students performed comparatively well on TIMSS, especially in science. This was a pattern not repeated, however, for the eighth graders. By the twelfth grade, on the tests of mathematical and scientific literacy, the USA students again performed poorly, although their showing was better in science than in mathematics. The key issue, of course, is the explanation of these findings, derived from the cross-sectional design of the study, especially for educationists and policy-makers.

The attempt to measure mathematical and scientific literacy with older students was also found in the TIMSS work. Orpwood (Chapter 3) evaluates the experience in an honest and compelling way, pointing out that the vision for this particular aspect of the overall study was so ambitious in conception that it was unlikely ever fully to be achieved. Like the Organisation for Economic Co-operation and Development (OECD) study discussed in the following chapter, these assessments of mathematical and scientific literacy were not intended to be curriculum-based but rather to 'ask whether school leavers could remember the mathematics and science they had been taught and therefore apply this knowledge to the challenges of life beyond school' (page 52). His account of the development process is revealing, emphasising as it does the inherent tensions between theoretical/academic debate and argument, political constraints and the practical need to produce assessments within a given time frame. Above all, he nicely summarises the constant competing demands of rich and varied questions and activities and the psychometric requirements of reliability. It is a situation known only too well by test developers.

Orpwood's conclusions are telling. He suggests that the fundamental and diverse purposes of the TIMSS work were never fully discussed, especially following the change of leadership and direction after 1993, and that without this clarification, the seeds of failure, or at least of only partial success, may well be sown. Were league tables to be the prime 'outputs' or was there a broader model, seeking to examine the broader influences that affect educational progress? His suggestion is that funding and political agendas need to be considered with great care since these have major implications for the focus and progress of the research.

The chapter by Andreas Schleicher (Chapter 4) explains the framework and approach of the new OECD international survey. He argues that this initiative is unique in several ways and therefore different from the IEA studies. The survey begins in the year 2000 and 32 countries will participate, of which 28 are OECD members. Its distinctive features include a regular three-year cycle of testing and a focus on evaluating the learning of

15 year-olds and assessing the cumulative yield of education systems at an age where schooling is largely complete. It will test the domains of reading literacy, mathematical literacy and scientific literacy, and within the three yearly cycles of testing, one of these will be the major focus (reading literacy in the year 2000) and the other two will have a more subsidiary role. This should provide regular monitoring of progress in these three fields in each cycle and more in-depth data on each every nine years.

The emphasis on this single, older age-group, more or less at the end of formal schooling in most societies, means that the surveys will not primarily be concerned with curriculum-specific content. Instead they will examine how well young people have acquired the 'wider knowledge and skills in these domains that they will need in adult life' (page 66). This may well serve to reduce the problems of curriculum commonalities and mismatches across countries that have so bedevilled the IEA surveys, although the tests will clearly need a content of some kind. To the extent that tests have to be about something, particularly in mathematics and science, this aim may not be achievable. Nevertheless, it will be a matter of great interest to see the assessment instruments and questionnaires, not to mention the results as they emerge.

Chapters 5 and 6 have a rather different character and pose rather different, yet related, questions. Elliot and Hufton (Chapter 5) report on some of the findings of a study carried out in England (in a small north-eastern city), Russia (in St Petersburg) and the USA (in Eastern Kentucky). These communities were chosen because economic hardship and decline and limited vocational futures seemed to be affecting the life chances of many of the students concerned. The study reports on student attitudes and reported behaviours and their role in explaining educational performance.

Interestingly, the results and discussion reported in a vivid way in this chapter may help to explain why students in England and the USA perform rather poorly in the IEA studies. The attitudes and behaviours of the samples of students in north-eastern England and in Kentucky seem 'inimical to the achievement of high educational standards' (page 109). Elliott and Hufton also argue that effort and ability, key variables an many studies, may be less important than familial, peer and cultural perceptions about the value of education. These are challenging findings and statements and they present considerable conceptual and methodological implications. This chapter serves to reinforce the point made above about the appropriateness and validity of encouraging a wide range of international comparative studies, not least for the new insights that smaller-scale studies can yield.

With the chapter by Reynolds, Creemers, Stringfield and Teddlie (Chapter 6) we return to large-scale international studies, but this time with a different kind of perspective. These researchers are heavily involved in school effectiveness research, but contextualise their work against a background of, for instance, the IEA studies. The project on which they report is the International School Effectiveness Research Project (ISERP), begun in 1991 after recognising both the common interests and limitations of strictly national investigations of the factors associated with effective schools. The results of such studies revealed some apparent contradictions: factors found to be significant in one country were not always influential in others. This suggested the need for wider theorising and more appropriate statistical modelling, potentially generating more 'complex and multi-faceted accounts' (page 118). The particular research questions focused on the factors that could be deemed general and those deemed more culturally specific in affecting student academic and social outcomes.

The main body of the chapter is concerned with what has been learned from the project, not in terms of results but in terms of conceptual and methodological insights, all highly pertinent in the context of this book. They report seven 'lessons', ranging from the importance of longitudinal, 'cohort studies', approaches to the need for classroom observation (however difficult in cross-national contexts) and for mixed research methods, to the vital importance of incorporating social as well as academic outcome measures (again, however, difficult). Compelling though these are, perhaps the most interesting lesson is the fact that there appear to be 'few agreed international constructs concerning effectiveness' (page 130): typically Anglo-Saxon notions seem to bear little relationship to those articulated in countries such as Taiwan or Hong Kong. This again reflects some of the points made earlier above. The authors' over-riding conclusion, however, is that whatever the difficulties, comparative education research is both important and beneficial. As they put it:

> 'There is a third reason for comparative investigation that is probably more important, concerning the possibility that within the right kind of comparative framework one can move beyond looking at the practices of other societies and actually so empathise with other societies that one can look back at one's own society with the benefit of their perspective' (page 133)

This 'acculturation' and use of another culture's 'lens' through which to view one's own is in line with other commentators' views about the benefits of international comparative studies. It is part of the vital self-evaluation process that the best international studies of education can yield, and it involves in part, as we saw earlier, a process of making explicit the implicit assumptions within all education systems. Comparative studies, well conceived and conducted, thus have the potential to increase our understanding in profound ways.

In Chapters 7 to 13, the emphasis of the book shifts towards a critical examination of some of the uses that have been, or might be made, of international comparisons of student achievement in science and mathematics. Much more is involved here than drawing over-hasty conclusions from so-called 'league tables' of relative performance. International comparisons such as TIMSS have generated a wealth of qualitative and quantitative data. Understanding how these data were collected and analysed is essential to any insight into any use that may be made of them. The problems of conducting international comparisons are formidable and they are well-known. Chapters 7 to 13 nonetheless offer clear evidence of the value of the data obtained and of the extent to which these problems can be overcome or accommodated.

Chapter 7 presents an overview of some of the uses that have been made of international comparisons of student achievement in science and mathematics and offers something of an introduction to the chapters which follow. The work of Angell, Kjaernsli and Lie, reported in Chapter 8, illustrates the rich potential of some of the TIMSS data as a source of information for teachers about students' thinking about a number of science topics. The methodology adopted by the authors to explore students' responses to free-response items in TIMSS is clearly capable of application to a much larger number of countries and its potential, thus far, remains under-exploited. In Chapter 9, Zuzovsky shows how international comparisons of student achievement can be used in association with national studies to generate data of considerable interest to educational policy makers. Writing with reference to Israel, Zuzovsky's approach can also be applied in other countries and contexts.

In Chapter 10, Kitchen offers a reminder of the need for caution when drawing conclusions or inferences from the results of international testing in mathematics. The reminder is both timely and important and it has a particular significance in the following chapter. This reports an attempt to compare the performance of students in England on both TIMSS and the tests which now form part of the national curriculum in that country. While direct comparison is impossible, interesting findings are uncovered and, as with the chapters by Angell *et al.* and Zuzovsky, the methodology is likely to be of wider interest.

In Chapter 12, Lokan explores the messages for mathematics education from TIMSS in Australia. Many of the weaknesses and strengths of school mathematics which she identifies will be familiar to those working in other countries and she draws attention to the on-going processes of curriculum reform and their relationship to the data generated by international comparisons.

The final chapter is concerned with the value of international comparative studies for a developing country. In this chapter, Howie captures vividly the organisational, practical, linguistic and other problems of administering TIMSS in the newly-democratic South Africa. Her conclusion is that, despite the severity of many of these problems, TIMSS – essentially a North American initiative – has much to offer science and mathematics education in the developing world. The wider message, that benefits can flow from international comparisons in spite of the problems associated with them, can serve as a fitting conclusion to the book as a whole.

ACKNOWLEDGEMENTS

We have explained above the origins of this book in a conference organised by the School of Education, University of Leeds. Our thanks must therefore go to all those who helped in the organisation of that event, including Janice Curry, Melanie Hargreaves and Bronwen Swinnerton. The work of Josie Brown was vital and special thanks must therefore go to her.

With regard to the preparation of the book, our thanks must first go to the contributors for their patience and willingness to respond promptly to our requests and queries. We hope they are satisfied with the results. Above all, however, the work of Sheila Mathison must be recognised. She has spent many hours preparing the manuscript and helping in the editing process and we owe her a considerable debt.

Finally, thanks are due to the publishing team at Kluwer and to the referees for their helpful and supportive comments. Any deficiencies that remain are our own.

Diane Shorrocks-Taylor
Edgar Jenkins

Chapter 1

International Comparisons of Pupil Performance
An Introduction and Discussion

Diane Shorrocks-Taylor
School of Education, University of Leeds, UK

1. THE BACKGROUND TO CURRENT INTERNATIONAL COMPARATIVE STUDIES OF STUDENT PERFORMANCE

Comparative studies in education are part of a long academic tradition dating back to the ancient Greeks and encompassing many different approaches. Postlethwaite (1988) suggests that these have ranged from nineteenth-century 'visiting' of other countries to see if they offered new insights about education, to the more systematic investigations of the twentieth century. These twentieth century studies have also varied, from the kinds of analyses that have sought to tease apart the complex relationships between education systems and wider economic, social and cultural factors to large-scale quantitative studies whose aim has been to answer questions about the effects of a range of influences upon educational achievement and improvement.

These traditions have not necessarily superseded each other, since present-day comparative education journals testify to a range and richness of comparative studies still being sustained. Small-scale, in-depth explorations can yield insights often missed (or not sought) in larger-scale enterprises, an end product that is justification in itself as well as providing potential hypotheses for wider investigation. This is a case powerfully argued by Howson (1998) in the context of comparative mathematics studies. Large-scale international studies are not, therefore, the end-point of an evolutionary tree, as it were, but represent an approach that may offer a distinctive

D. Shorrocks-Taylor and E.W. Jenkins (eds.), Learning from Others, 13–27.
© 2000 *Kluwer Academic Publishers. Printed in the Netherlands.*

contribution to our understanding, complementing the insights generated by other kinds of approach and other kinds of questions.

However, since the late 1950s, with the founding of the International Association for the Evaluation of Educational Achievement (IEA), such large-scale comparative studies have taken on considerable significance in education. Their reported findings seem to occupy educationists' and policy makers' minds to a disproportionate degree. In the case of the IEA studies this is especially true. These studies represent a collaboration of over 45 countries and their aim has been to conduct comparative studies focusing on educational policies and practices in order to enhance learning within and across systems of education. From the beginning, the IEA has been committed to studying learning in the basic school subjects and to conducting, on a regular basis, surveys of educational achievement. Over time, these 'outcome' data have been increasingly linked to analyses of the effects of curriculum and school organisation upon learning and the relationship between achievement and pupil attitudes etc.

The IEA is organised in a broadly 'collegial' way, with delegates from member countries/institutions attending an annual General Assembly. When international projects are conceived, they are discussed widely, involve multi-national Steering Groups and have nominated National Project Co-ordinators from all participating countries. Each national institution is expected to have not only research expertise but also direct links both with schools of all types in the country and with policy makers. The funding derives from participating governments, which often means that the poorest nations cannot take part in the study activities.

This is not to suggest, however, that the IEA is the only large-scale player in the international comparative education field. The International Assessment of Educational Progress (IAEP) also figures large and has conducted surveys of mathematics and science performance since the late 1980s. However, it is closely linked to the Educational Testing Service (ETS) in the USA, and financial responsibility is borne by the Service. This gives a rather different character to the studies.

The Organisation for Economic Co-operation and Development (OECD) is increasingly figuring in international studies too. In collaboration with its Centre for Educational Research and Innovation (CERI), it has for many years published its OECD Indicators which summarise comparative educational statistics in relation to its member nations (29 at the last count). These cover areas such as educational expenditure, participation in formal education, the learning environment and the organisation of schools and (in collaboration with the IEA) student achievement. As Chapter 4 in this book indicates, the OECD has recently moved directly into the field of comparative studies of student performance through a new project implemented in 1999 and it is interesting to compare the rationale and methods of the two surveys. However, to date, the IEA studies are most representative of the *genre* of international comparisons of student performance, so it is to these we turn for lessons about the pros and cons of the endeavour.

The first IEA study in 1959 was on elementary and secondary mathematics education, carried out in 13 countries. Since then other studies have surveyed, in addition to mathematics, science, reading comprehension, literature, French and English foreign languages, written composition, computers in education and more, well summarised in Postlethwaite (1987) and Goldstein (1996).

However, perhaps the most well known pieces of work carried out under its auspices are the large-scale surveys of mathematics and science achievement. The first of these was in the early 1960s in mathematics (FIMS – the First International Mathematics Study), followed by the equivalent science survey (FISS – the First International Science Study) in the early 1970s. The second round of surveys were conducted in 1982-83 (SIMS – the Second International Mathematics Study) and in 1984 (SISS – the Second International Science Study). Most recently, of course, has been the Third International Mathematics and Science Study (TIMSS) focusing on both mathematics and science. This is one of the focus points for this book and will be considered again in more detail in this chapter and throughout most of the rest.

What is interesting in considering such a catalogue of studies is the fact that over the years, not only has the range of subjects broadened but the contextualising of the results has been gradually extended, providing more details about such dimensions as curriculum, school organisation, and background information about the students (Goldstein, 1996). As Howson (1998) points out, there has been a considerable learning curve over the years of the IEA surveys.

The number of participating countries has also increased which raises important questions about the motivation of the participants and what they hope to gain from such participation.

It could be argued that countries might be better served by investigating and understanding their own educational systems in more detail rather than looking to international surveys to yield useful information. This is especially the case when the problems and constraints of such studies are taken into account, a topic that will be dealt with in the later sections of this chapter. So why do countries participate and what do they hope to gain from the enterprise? This is the major focus of the chapters later in this book, but it is worth outlining some preliminary points here, by way of introduction.

Some insights into this question of national motivation have been provided by various authors and they largely emphasise the different kind of information and understanding that international studies provide about national systems. Such studies, if well conceived and conducted, can provide information about the nature and quality of their own system that can contribute to the more realistic evaluation and monitoring of schooling and its quality. Comparative studies can also be helpful in the broader and better understanding of the reasons for the differences in outcomes. By considering outcomes across different countries, better understanding of the effects of such factors as teacher training, time spent in school, parental involvement etc. can be generated and be of use in policy-making in the national context.

In 1992, when the TIMSS study was being put in place, two editions of the journal *Prospects* were dedicated to an analysis of different aspects of the work, including the reasons for participation by several countries. It is interesting to consider the points made in three of the papers from very contrasting countries, namely a developed western nation (the USA) a rather different but nevertheless technologically advanced country (Japan) and a developing country (Kuwait).

Griffith and Medrich (1992) began by pointing out that there was widespread interest in the results of international surveys by educationists and policy makers in the USA, not least because of the direct link perceived between the quality of the education system and economic competitiveness. The economic and educational performance (at that time) of countries in the far east seemed to suggest that a well-educated workforce was closely linked with progress in the economy and increased output. The 1980s and early 1990s had witnessed a sense of apprehension in the USA about educational policies, resulting in the setting of national educational goals. In the context of a highly decentralised educational system, this represented a major step, driven by the pervasive sense that all was not well in American schools. The argument made by these authors was that measuring and evaluating success in the education system required both national and international perspectives and a move from inward-looking and insular attitudes towards greater openness to new ideas and ways of doing things. Self-improvement relies in part upon learning from others.

The final part of their argument however (Griffith and Medrich, *op cit.*) re-emphasises the view that if international comparative studies have this significance, then it is imperative that the data upon which the outcomes are based must be of the highest possible quality, placing very considerable burdens on those devising and administering the studies. Persistent rumblings about potential problems in the sampling, response rates, marking and coding of the data serve only to undermine confidence in the results.

One final point made about American perceptions of the value of participation is that, in the USA, policy makers increasingly ask questions not about the outcomes but about systems and in particular, the effects of such contextualising factors such as student attitudes, the role of parents and the effects of housing, health etc. The focus, it is argued, is now on the correlates of high quality school performance and the issue of what makes a difference.

From a rather different perspective, Watanabe (1992) argues that what Japan wants is some indication of how successful it is in producing 'useful and productive adults' (*ibid.,* page 456-462). More particularly, Japan participates for the following reasons.

– The IEA surveys are co-operatively designed and this yields important insights in both research and curriculum terms. One example, is the conceptualisation of the curriculum in the IEA studies, namely the *intended curriculum,* the *implemented curriculum* and the *achieved curriculum,* an approach and way of thinking not previously considered in Japanese education.

– Past experience has shown that participation provides experiences that can have a direct and positive impact on policy and practice, for example in the use of computers in schools.

– The fact that each country can to some extent adjust the survey or participate in only aspects relevant to its own needs, also increases the potential to develop skills and strategies likely to be of national use, for example the testing of practical or 'applied' skills in science and mathematics. In Japan this has led to a change from a more factual and recall-based science curriculum to a more inquiry-based one.

– The fact that contextualising information has been increasingly sought and applied in the analyses means that factors affecting learning outcomes (resources, teaching approaches, curriculum organisation and so on) can potentially be better understood.

– Participation means that those involved learn new research skills in study design, administration, analysis and interpretation that can transfer directly into the national system.

Watanabe's conclusion is that participation in international surveys has a direct benefit to the Japanese national system and has led to further regional as well as international collaboration. Participation requires discussion and compromise which in itself may be important in an international context. Most importantly, however, it provides an antidote to attitudes that try to suggest that there are easy answers to difficult educational problems throughout the world or miracle solutions that can be applied irrespective of the kind of education system and its socio-cultural context.

Hussein (1992) brings the perspective of a developing country (Kuwait) with very different characteristics and constraints. He paints a picture of an education system where many new initiatives come and go in what appears to be a quite arbitrary manner, and where the evaluation of the outcomes and effects is seldom carried out in an objective way. Small and powerful groups within the system make judgements, not necessarily with the benefit of evidence. In this situation, international comparative studies offer the possibility of providing some kind of external benchmarking as well as new ideas in the curriculum and teaching approaches. Such comparisons offer one of the few sources of valid evidence about the comparative success or failure of the school system in Kuwait. He argues that participation has already yielded benefits in relation to promoting assessment research, survey technologies, the national monitoring of achievement and the interpretation of data. It has also provided a framework and approach which can be applied in the local/national context.

These three rather different perspectives are illuminating and no doubt this array of motives would be even wider with a broader sample of information from other countries. In general terms, they all suggest that learning from each other is an important gain, and that the process of participation requires self-evaluation which in turn may lead to assumptions being questioned and what was previously understood only implicitly now being made explicit and so examined in a more critical way. The developing countries also seem to emphasise the importance of obtaining important information about the performance of their own students and the benefits the training provides in survey techniques, research design, implementation and data analysis.

What the three papers reviewed here also suggest, however, is that countries also have their own more particular motives for participating, over and above these commonly cited ones. Motives and agendas vary but it seems that international surveys, at their best, can meet a wide range of perceived needs on the part of those participating and can encourage a sharing of perspectives and a forum for the discussion of educational ideas that can serve many purposes.

These may appear somewhat naïve statements, which need to be counterbalanced by the realities of participation – the time constraints, power struggles and practical and technical problems that also arise. It is significant that Griffith and Medrich insert the major caveat about the quality of the information that goes into producing the outcomes. It is to the constraints and problems that we now turn.

2.   THE PROBLEMS IN INTERNATIONAL COMPARATIVE STUDIES

In a sense, the whole of this book provides an evaluation of the strengths and weaknesses of international comparisons, so the issues should not be pre-judged here. However, as a starting point, it is helpful to begin to draw together some of the more general comments made about the limitations and constraints of such studies.

It is often stated that international surveys embody six basic stages or dimensions (Loxley, 1992). These are:
- the conceptual framework and research questions;
- the design and methodology of the studies;
- the sampling strategy;
- the design of the instruments;
- data collection, processing and management;
- the analysis and reporting of the findings.

Taken together, these will provide a framework for summarising the constraints and problems.

2.1   The conceptual framework and research questions in international studies

As in any piece of research, the starting point is the set of fundamental questions to be posed and answered. This is clearly the case, irrespective of the scale of any study, since all design, methodology and analysis decisions flow from this basic source. The questions may be framed in different ways, with different degrees of detail and at different levels of generality and they may be stated as questions or hypotheses. An interesting comment made by Howson (1998) is that most large-scale studies (the IEA ones for example) have worked within an inductive framework, analysing situations and contexts, collecting data and analysing the results (mostly in terms of rankings, percentiles, and correlation) generating some hypotheses along the way which the data can help to illuminate. Seldom has a hypothetico-deductive model been used, in which specific hypotheses are tested, which from the beginning determine the whole design, analysis and reporting.

The corollary of this is that the products of the comparisons, the rankings or league-tables take on a competitive edge. In fairness, the IEA studies claim to have tried to discourage this international game of 'who can come top of the list', a kind of intellectual World Cup, but the country rankings are often the first focus point for politicians and policy makers. There are parallels here with the national situation in England, where the introduction of a National Curriculum and national testing system at several age points throughout schooling has lead to the spectator sport of league-table watching and league-table bashing. Raw scores for all schools (primary and secondary) are published in both national and local newspapers and schools judged accordingly. Parents then frequently vote with their feet. Critics of this procedure point out the huge oversimplification and unfairness of publishing raw results, unadjusted for type of school and socio-economic location, where contextualising information that would allow the more valid interpretation of the outcomes is not included.

It is probably true that the early international comparative studies were open to more criticism than later ones on this score, since what is noticeable about many later studies is the attempt to contextualise the outcomes by seeking much more information about school organisation, curricula, practice in schools and the attitudes of students and their families alike. It is clear (see Chapter 2) that the recent TIMSS studies have tried to address this question in relation to its conceptual framework. One of the earlier models used in the IEA studies shows the range of factors to be taken into account.

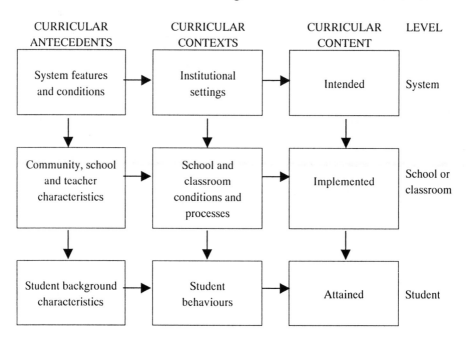

Figure 1.1. An IEA research model

In this kind of model, the rows indicate the levels of data collection in order to be able to provide information on between-system differences, between-school/class differences, and between-student differences. Meaningful international comparisons can only be made if there is sufficient common ground among participating countries in terms of the intended, implemented and received curricula. In the absence of total common ground (which there clearly is not) the curriculum within each country is reflected in the achievement tests in a very approximate kind of way, essentially using a 'best-fit' kind of approach with all its attendant non-congruencies.

However, we need also not just to relate educational achievements to these wider influences but also to the institutions and values of the society and to a model of how educational systems as a whole work and reflect curriculum based learning. In an international context, the model also needs to be sufficiently general to measure the global effects of schooling on student learning. We are still far from understanding the links between the curriculum as intended, the curriculum as implemented and the curriculum as received by pupils, particularly in such a wide range of socio-cultural and pedagogic contexts.

One final point on the matter of the fundamental research questions of international studies concerns the political dimension that clearly comes into play in surveys funded from government sources. As Goldstein (1996) points out, the plus side of such an arrangement is that government support is guaranteed, as is access to schools: the down-side is that governments may pursue narrower political agendas than a purely research-led initiative might allow.

International surveys of the kind mostly focused on in this book have investigated a wide range of subjects, from mathematics and science to reading comprehension, literature and civic education. To some extent, this therefore gives the lie to critics who argue that they measure only limited kinds of learning and moreover those which are easiest to measure through straightforward pencil-and-paper tests or simple activities. It has to be conceded however, that there have been few attempts to compare creative and aesthetic qualities in pupils (Morris, 1998). The conceptualisation of most studies to date has been in terms of traditional school subjects.

As we have seen, such studies have frequently sprung from an apprehension felt in many countries that all is not well in education systems and that it ought to be possible to improve schooling, with its supposed economic benefits, in an international context of shared discussion and exploration of organisation and practice. Learning from each other is the rhetoric even if the reality distils into little less than comparing results in a league-table kind of way.

However, one of the most pervasive negative comments made about international surveys is that not only are the comparisons themselves problematic, but so is the notion that it is ever possible to take an idea or experience generated in one culture and somehow transpose this to another. The case for international comparisons therefore seems flawed on both theoretical and practical grounds. Above all, education systems must be understood at a macro-level within the wider social, economic and political systems, and the cultural values embodied within these.   Given this complexity and embeddedness, it seems highly unlikely that practice or policy from one system would have easy messages or possibilities for other systems.

3.   THE DESIGN AND METHODOLOGY OF THE STUDIES

Design and methodology obviously follow on from the basic research questions and from a recognition of the time and resource constraints of any study.   Two interesting points emerge from an overview of the more recent international surveys: on the whole they focus on very similar age-groups and in the main they utilise a cross-sectional design.

The focus of the major studies has been principally on one elementary age-group (usually 9 to 10 year-olds) and one secondary age-group (usually 13 to 14 year-olds).   Some studies have also focused on the secondary age-group closest to the end of schooling (often 15 to 16 year-olds).   The choice of 9 year-olds has often been dictated by the commencement age for schooling.   To choose younger age-groups could mean that some pupils in the sample had only recently started school in some countries, a selection decision that would not support the aim of evaluating the effects of schooling, unless it was used in a 'baseline' kind of way.   It is interesting, however, that students of age 11 or 12, a common age for changing phases of education in many countries, have seldom been investigated.

The issue of cross-sectional versus longitudinal designs is an important one.   Virtually all studies so far have adopted a cross-sectional design, where two or more age-groups are included at the same point in time, with the samples being entirely separate.   This allows results to be generated which describe learning outcomes at different ages and, if this is repeated over time, as with the major IEA studies and the new OECD survey, a certain amount of time-trend data can emerge.   However, what this kind of data cannot furnish is any real insight into the factors affecting learning, and their long-term influence.   Correlational analyses of cross-sectional data can suggest relationships between a range of variables but real explanatory power can only come from longitudinal investigations.

It is for good reasons, however, that few longitudinal studies have been carried out in international contexts.   Such studies are infinitely more difficult to design and implement, not least because of retaining contact with respondents over long periods of time.   Attrition rates are frequently high, a fact that needs to be built into the design and sampling from the beginning, and if it is difficult to retain samples in more economically advanced countries, how much more difficult is it perhaps in remote parts of less developed societies?

4. THE SAMPLING STRATEGY FOR THE STUDIES

The sampling unit in most studies is the school, and the sampling strategy often involves stratification by region, type or size. It should be pointed out that stratification by region may not be problematic in smaller, more developed countries but may present considerable difficulties in large, less developed countries. Using the school as the unit of selection works well for younger age-groups of students but frequently presents difficulties with older ones, for instance those at the point of leaving school who may be in a range of institutions other than schools, perhaps in vocational colleges, day-release in work etc.

However, even for the younger age-groups, the perennial problem of age selection or grade selection looms large. To select by Grade or Year Group of schooling allows the impact of the cumulative effects of schooling to be measured, but there is enormous variation in the way Grades/Year groups are made up in schools across the world. In some countries, the Grades bear a close relationship to chronological age in that a single age is more or less represented by a single Grade or school Year Group. But in other situations, the range of chronological ages in a single Grade can be enormous, largely as the result of students repeating grades until they are successful or students beginning their schooling at different ages but at the same grade.

Age-based sampling seems better for comparability, but children begin school at different ages in different countries so it is by no means straightforward. The IEA studies have overcome the problem by including both kinds of sampling: pairs of Grades around the selected age points have been extensively used.

Sampling theory suggests that the most important factor in creating a representative and defensible sample is valid information about the characteristics and make-up of schools in the potential sample. This sounds trite, but it is clearly a problem in some countries where information systems may not be well established. As Ross (1992) argues, an important aspect in school sampling is knowing about the homogeneity or heterogeneity of the students attending each school. If schools are very homogeneous in intake (the result of selection and streaming, for instance) then more will need to be sampled than if the schools are more heterogeneous in character (i.e. with a more comprehensive intake).

Finally, there is the problem of response rates. The quality of the final results of any comparative study will depend closely upon the quality of the sample selected and the quality and representativeness of the achieved sample. Poor response rates can distort and bias the outcomes in significant ways, and yet getting schools and students to complete the measures and instruments is a major problem in developed and less developed countries alike.

5.   THE DESIGN OF THE INSTRUMENTS IN INTERNATIONAL STUDIES

To some extent this has already been touched upon in earlier sections, in discussions of school curricula and how these may or may not relate to the test items used internationally. But the issue does not just concern the content of the testing itself.

The recent studies (for example TIMSS and the new OECD initiative) have sought, quite justifiably, to include such dimensions as student background information, classroom observations, and the assessment of practical work within their remit. Laudable though this may be in design and outcome terms, it nevertheless introduces many more problems as well as possibilities into the dataset. For example, Goldstein (1996) notes that a question such as 'number of children in the family' can be deceptively simple. In some countries, the definition of 'family' may be very different and respondents may interpret the question in different ways. Should adult brothers or sisters be included even if they are no longer living at home and should step-siblings be counted? The cultural connotations are significant and the question itself reveals a considerable cultural imperialism. Even more dangerous ground is revealed by questions intended to access some measure of the social background and status of participating students.

The observation of classrooms and of practical teaching activities also has its pitfalls. Schmidt *et al.* (1996) provide a telling account of the amount of preliminary discussion time needed among representatives of the six participating countries in the investigation of mathematics and science teaching, taking part in the TIMSS exercise. They suggest (page 7) that:

'English was the common language used for the meetings. Even so, it soon became apparent that the same language was not being spoken at all. The problem was not simply one of translation from French or Norwegian or of limited English skills by non-US team members. It became clear that operational definitions for common educational expressions differed among countries. It also became clear that much was not in common among the participating countries in thinking about education. Even the behaviours or events thought important in characterising instructional practices were found to differ among the countries'.

Finally in this section, the question of translation must be raised, translation of both the test questions and the other instruments used in the international surveys. The first versions of the instruments may be in one language but these have to be translated into the languages of the other participants. The overwhelming judgement about such processes is that they are difficult, complex and subtle. Using bilinguals and back translation may go some way to solving the problem but translation is seldom direct, especially across languages that are structurally and semantically very different. Because of this, the term 'translation' is often replaced by 'adaptation' since this is closer to the kind of process that has to take place.

On a rather different level, most of the preparatory discussions in international studies takes place in English, used as a sort of *lingua franca*. This may mean that less than fluent English speakers who are representing their countries may be able to make less of an input and impact than they perhaps could and should.

6. DATA COLLECTION, PROCESSING AND MANAGEMENT IN LARGE INTERNATIONAL STUDIES

Needless to say, computers have transformed the processes of data collection, processing and management in large-scale international studies. But this may be a double-edged sword. There may be a temptation to over-specify and over-collect data knowing that computers can handle the work, which may in turn lead to being swamped by data that may not be used or usable. Schleicher (1992) outlines some of the problems and suggests that at the point of data collection, failure can occur for these over-collection reasons, or as a result of poor training and project management. The ideal study begins from the main research questions and collects only as much data as are dictated by these. Similarly, the analyses should be specified from the beginning so that all data needs are known and taken into account from the earliest stages.

As in any research, best practice suggests that quality and standards checks need to be built in throughout the study so as to eliminate, as far as possible, errors in printing, coding, data entry and analysis. However, in international surveys these problems are writ large indeed.

7.   THE ANALYSIS AND REPORTING
        OF THE FINDINGS

In reporting findings, it is the case that simplicity is the ideal. End-users should be able to pick up the major findings and messages in a concise way, again arguing for clear initial questions that can be answered in comprehensible ways. However, this is not always easy to achieve, particularly if arguments and relationships between variables are complex. The problem with simplicity is also that it may lead to the kinds of oversimplifications referred to earlier, encouraging a focus on, for instance, tables of rankings rather than more useful and interesting findings. One way of guarding against this is to report the findings in a disaggregated way, for instance by topic area within a subject, or by the kinds of cognitive skills displayed (recall, application, evaluation etc.).

Many of the larger studies reported here are repeats of similar earlier studies and they may even include some common questions from one administration to another. In this case, data on trends over time may be possible but, as Goldstein (1996) suggests, these are very difficult to measure and interpret. If performance on an item changes, is it the performance of the population that has changed or is it the item or is it both? The same item may take on different significance from administration to administration, because of curricular changes or wider societal influences so the effects may be hard to disentangle. Goldstein (*op cit.*) argues that the only solution is a multi-level modelling approach which may allow more appropriate and convincing results to be generated.

8.   SOME CONCLUSIONS

A few key points emerge from this catalogue of the potential limitations and problems of international comparative studies. The first is that from the early days when the first mathematics studies took place, conceptualisations, designs, implementation procedures and data analyses have developed and improved greatly. The earliest IEA studies (for example the First International Mathematics Study) were heavily criticised for their lack of attention to the significance of the differing curricula in the participating countries, their poor contextualisation of the outcomes and the heavy reliance on multiple-choice question formats (Freudenthal, 1975). Subsequent studies have sought to rectify these shortcomings: certainly they have proved much more successful in terms of data handling and analysis. Reports are now delivered fairly rapidly after final data collection and can therefore potentially have greater impact at both national and international levels.

The summary has also shown that there are still fundamental issues and problems inherent in such studies, only some of which have been or are able to be addressed. The judgement is whether there is more to gain from carrying out such investigations, even with their flaws, than in not even attempting to explore the approach further and slowly extend and improve them. It is a national and international cost-benefit analysis that needs to take place.

REFERENCES

Goldstein, H. (1996) International Comparisons of student achievement, in Little, A. and Wolfe, A. (eds), *Assessment in Transition: learning, monitoring and selection in international perspective* (London: Pergamon)

Griffith, J. E., and Medrich, E. A. (1992) What does the United States want to learn from international comparative studies in education?, *Prospects,* 22, pp. 476-485

Freudenthal, H. (1975) Pupils' achievement internationally compared – the IEA, *Educational Studies in mathematics, 6,* pp. 127-186

Howson, G. (1998) The value of comparative studies, in Kaiser, G., Luna, E., and Huntley, I (eds) *International Comparisons in Mathematics Education* (London: Falmer Press)

Hussein, M. G. (1992) What does Kuwait want to learn from the Third International Mathematics and Science Study (TIMSS)?, *Prospects, 22,* pp. 463-468

Loxley, W (1992) Introduction to special volume, *Prospects*, 22, pp. 275-277

Morris, P. (1998) Comparative education and educational reform: beware of prophets returning from the Far East, *Education* 3-13, pp. 3-7

Postlethwaite, T. N. (1988) *The Encyclopaedia of Comparative Education and National Systems of Education*, Preface (Oxford: Pergamon Press)

Ross, K. N. (1992) Sample design for international studies, *Prospects*, 22 pp. 305-315

Schleicher, A. and Umar, J. (1992) Data management in educational survey research, *Prospects,* 22, pp. 318-325

Schmidt, W. H., Jorde, D., Cogan, L. S., Barrier, E., Gonzalo, I., Moser, U., Shimizu, Y., Sawada, T., Valverde, G. A., McKnight, C. C., Prawat, R., Wiley, D. E., Raizen, S. A., Britton, E. D., Wolfe, R. G. (1996) *Characterising Pedagogical Flow: An Investigation of Mathematics and Science teaching in Six Countries*, p. 229 (Dordrecht: Kluwer)

Watanabe,R. (1992) How Japan makes use of international educational survey research, *Prospects*, 22, pp. 456-462

Chapter 2

International Comparisons of Student Achievement
Perspectives from the TIMSS International Study Center

Michael O Martin and Ina V S Mullis
TIMSS International Study Center, Boston College, USA

1. THE VISION

In the late 1980s, researchers from around the world came together to plan the International Association for the Evaluation of Educational Achievement's (IEA) largest and most ambitious study to date, the Third International Mathematics and Science Study (TIMSS).[2.1] From the early stages of the project, TIMSS was to be a comprehensive international study of mathematics and science education that would address student achievement as well as the explanatory factors associated with achievement. The planners of TIMSS aimed to design a study that would surpass all previous comparative education studies in size and scope and provide a more in-depth view of mathematics and science education around the world. With two subject areas and target populations at three critical points in students' schooling, TIMSS has accomplished that.

Since its inception in 1959, the IEA has conducted a series of international comparative studies designed to provide policy makers, educators, researchers, and practitioners with information about educational achievement and learning contexts. IEA studies have traditionally focused on the output of educational systems, that is, the attitudes and educational achievements of students, and attempted to relate these outputs to the inputs which were antecedent to them. The purpose has been to learn more about

---

[2.1] Funding for the international co-ordination of TIMSS was provided by the US Department of Education's National Centre for Education Statistics, the US National Science Foundation, and the Applied Research Branch of the Strategic Policy Group of the Canadian Ministry of Human Resources Development

D. Shorrocks-Taylor and E.W. Jenkins (eds.), Learning from Others, 29–47.

the factors which influence student attitudes and achievement, and which can be manipulated to bring about improvements in attitudes and achievement, or efficiencies in the educational enterprise. IEA studies are traditionally collaborative ventures among educational systems around the world that come together to plan and carry out the studies. An international centre directs the international activities of a study, and a representative in each country directs the national implementation of the study. TIMSS also was to be carried out in a collaborative manner, under the direction of an international study centre.[2.2]

The designers of TIMSS chose to focus on curriculum as a broad explanatory factor underlying student achievement. From that perspective, curriculum was considered to have three manifestations: what society would like to see taught (the intended curriculum), what is actually taught in the classroom (the implemented curriculum), and what the students learn (the attained curriculum).

Drawing on the tripartite model of curriculum and the notion of educational opportunity, four major research questions were posed and used to build the study and the TIMSS data collection instruments:
− What are students expected to learn?
− Who provides the instruction?
− How is instruction organised?
− What have students learned?

1.1   What are students expected to learn?

This question was investigated through the analysis of curriculum guides and textbooks, which were considered the major vehicles through which society's expectations for student learning, that is, the intended curriculum, are imparted and through questionnaires about the educational systems which were to be administered to experts in each country. To provide a framework for conducting the curriculum analysis and developing the TIMSS data collection instruments, the TIMSS curriculum frameworks were developed early in the study as an organising structure within which the elements of school mathematics and science could be described, categorised, and discussed. In the TIMSS curriculum analysis, the frameworks provided the system of categories by which the contents of textbooks and curriculum guides were coded and analysed. The same system of categories was used to collect information from teachers about what mathematics and science

---

[2.2] From 1990 to 1993, TIMSS was directed by the International Co-ordinating Centre at the University of British Columbia, Canada. From 1993 to the present, TIMSS has been directed by the TIMSS International Study Centre in Boston College's School of Education, the United States.

students have been taught. Finally, the frameworks formed a basis for constructing the TIMSS achievement tests.

Subject-matter content, performance expectations, and perspectives constitute the three dimensions, or aspects, of the TIMSS curriculum frameworks. *Subject matter content* refers simply to the content of the mathematics or science curriculum unit or test item under consideration. *Performance expectations* describe, in a non-hierarchical way, the many kinds of performance or behaviour that a given test item or curriculum unit might elicit from students. The *perspectives* aspect is relevant to analysis of documents such as textbooks, and is intended to permit the categorisation of curricular components according to the nature of the discipline as reflected in the material, or the context within which the material is presented. Each of the three aspects is partitioned into a number of categories, which are themselves partitioned into subcategories, which are further partitioned as necessary.

TIMSS Science Frameworks	TIMSS Mathematics Frameworks
Content • Earth Sciences • Life Sciences • Physical Sciences • Science, technology, and mathematics • History of science and technology • Environmental issues • Nature of science • Science and other disciplines **Performance Expectations** • Understanding • Theorizing, analyzing, and solving problems • Using tools, routine procedures • Investigating the natural world • Communicating **Perspectives** • Attitudes • Careers • Participation • Increasing interest • Safety • Habits of mind	**Content** • Numbers • Measurement • Geometry • Proportionality • Functions, relations and equations • Data representaiton, probability and statistics **Performance Expectations** • Knowing • Using routine procedures • Investigating and problem solving • Mathematical reasoning • Communicating **Perspectives** • Attitudes • Careers • Participation • Increasing interest • Habits of mind

Figure 2.1. Major Categories of Curriculum Frameworks

1.2   Who provides the instruction? And how is instruction organised?

These two questions were addressed through detailed questionnaires administered to mathematics and science teachers of the students tested for TIMSS, principals of the students' schools, and to the sampled students. The questionnaires sought information about teachers' characteristics, education, the teaching environment, subject-matter orientation, and instructional activities, with the aim of providing information about how the teaching and learning of mathematics and science are realised, that is, the implemented curriculum.

In conjunction with the activities described above, the United States sponsored two additional parts of TIMSS, which were carried out at the eighth grade in Germany, Japan, and the United States. Videotapes were made of mathematics lessons in randomly selected classrooms that participated in the main study. Teachers were filmed teaching a typical lesson, and these tapes were analysed to compare teaching techniques and the quality of instruction. Also, ethnographic studies were conducted in these three countries to study education standards, methods of dealing with individual differences, the lives and working conditions of teachers, and the role of school in adolescents' lives.

1.3   What have students learned?

This question was addressed through a large-scale assessment of student achievement in mathematics and science. TIMSS chose to study student achievement at three points in the educational process: at the earliest point at which most children are considered old enough to respond to written test questions (Population 1 – third and fourth grades in most countries); at a point at which students in most countries have finished primary education and are beginning secondary education (Population 2 – seventh and eighth grades in most countries); and the final year of secondary education (Population 3 – twelfth grade in many countries).

The aim of the developers of TIMSS was to test students in a wide range of mathematics and science topics through a range of question types. Recent IEA studies, particularly the Second International Mathematics Study, placed great emphasis on the role of curriculum in all its manifestations in the achievement of students. This concern with curriculum coverage, together with the desire of curriculum specialists and educators generally to ensure that both subjects be assessed as widely as possible, led to pressure for very ambitious coverage in the TIMSS achievement tests. Further, there was concern that the assessment of student knowledge and abilities be as 'authentic' as possible, with the questions asked and the problems posed in a form that students are used to in their everyday school experience. In particular, there was a requirement that test items make use of a variety of task types and response formats, and not be exclusively multiple-choice. In addition, the inclusion of a performance assessment was viewed as critical to the design of TIMSS. Finally, recognising that the outcomes of schooling include the affective domain, the student questionnaire was intended to measure students' attitudes towards mathematics and science.

2. THE CHALLENGES

2.1 Resources and time

In designing and implementing the study, TIMSS faced the usual constraints and challenges of an ambitious undertaking that relies on people, time, and funding. The designers and management staff of TIMSS had to consider the levels of funding available to manage the study at the international level and to implement the study in each country. Time was another constraint. The time for the development of the study, while relatively lengthy, was not unlimited. The main test administration was to take place in 1994-95, which required that all instruments be developed, field-tested, and ready to be administered by that time, and that all procedures be developed and understood by participants. Another time constraint was the amount of time that could reasonably be asked of respondents – students, teachers, and schools. For example, the participating countries agreed that they could not reasonably require more than 70 minutes of a primary-school student's time, or 90 minutes of a middle-school or final-year student's time for the test administration.

Reconciling these time constraints with the ambitious coverage goals of the TIMSS achievement tests was a lengthy and difficult process. It involved extensive consensus-building through which the concerns of all interested parties had to be balanced in the interests of producing a reliable measuring instrument that could serve as a valid index of student achievement in mathematics and science in all of the participating countries. The tests that finally emerged, which were endorsed by all participating countries, were necessarily a compromise between what might have been attempted in an ideal world of infinite time and resources, and the real world of short timelines and limited resources. The background questionnaires that were developed for students, teachers, and schools necessarily were constrained by the time that could be asked of the individual respondents and thus could not address every single question of interest to each constituent.

2.2   Methodological challenges

The design and development of TIMSS was a complex process involving representatives from the participating countries, technical advisors, mathematics and science educators and subject matter specialists, and the major funding agencies. Not surprisingly, the involvement of many constituents meant that there were many competing priorities, requiring innovative methodological solutions and involving many trade-offs and compromises before reaching the final design. Among the major components of TIMSS that posed methodological challenges were the curriculum analysis and the achievement testing and estimation of proficiency.

The curriculum analysis was an innovative attempt at a comprehensive analysis of curricular materials in each participating country, to provide an objective characterisation of the intended curriculum. Such an attempt at a comprehensive analysis had never been conducted on the scale planned for TIMSS. To conduct the curriculum analysis, methods of sampling, coding, and analysing curricular materials had to be developed, implemented, and reported, which posed a daunting challenge for the staff at Michigan State University, who were responsible for this component of TIMSS.

A second major methodological challenge was the measurement of student achievement, which is always a challenge under any circumstances. The measurement of student achievement in two subjects at three student levels in 45 countries (through the local language of instruction), in a manner that does justice to the curriculum to which the students have been exposed and that allows the students to display the full range of their knowledge and abilities, is indeed a formidable task. Since there is a limit to the amount of student testing time that may reasonably be requested, assessing student achievement in two subjects simultaneously constrains the number of questions that may be asked, and therefore limits the amount of information that may be collected from any one student. Matrix sampling provided a solution to this problem by assigning subsets of items to individual students in such a way as to produce reliable estimates of the performance of the population on all the items, even though no student has responded to the entire item pool. The TIMSS test design uses a variant of matrix sampling to map the mathematics and science item pool into eight student booklets for Population 1 and Population 2, and nine booklets for Population 3.

The demands for maximum content coverage translated into challenges in reporting achievement. TIMSS relied on item response theory (IRT) scaling to summarise the achievement data. Specifically, the scaling model used in TIMSS was a multidimensional random coefficients logit model.[2.3]

[2.3] See the *TIMSS Technical Report, Volume II*, Chapter 7 by Adams, Wu, and Macaskill for an explanation of the TIMSS scaling procedures

And, because each student responded to relatively few items within each subject matter content area, it was necessary to use multiple imputation, or 'plausible value', methodology to estimate student proficiency reliably.

2.3 Quality Assurance

In addition to the usual constraints facing a research endeavour of the size and scope of TIMSS were constraints that grew out of the recent interest in international studies of student achievement. More and more, policy makers are turning to international comparisons to gauge the status of their education systems. The increased visibility of international studies as an information source and a vehicle for reform has led to demands that data collected across countries be internationally comparable and of the highest quality possible, a burden not placed upon earlier international studies of student achievement. TIMSS responded to these demands with a program of quality assurance that paid particular attention to ensuring that the national samples were as comparable as possible and fully documented, that the data collection procedures were standardised across countries, and that the data were put through an exhaustive system of checks and balances.

3. THE REALITY: THE U.S. EXPERIENCE IN TIMSS

Despite the challenges associated with designing and implementing such a large and ambitious project, TIMSS was quite successful in meeting the original aims of the study. Taken together, TIMSS included many different parts: system descriptions, achievement tests, questionnaires, curriculum analyses, videotapes of mathematics classroom instruction, and case studies of policy topics. The use of multiple methodologies, each of which provides an important advance in its field, provides an unprecedented opportunity for countries to understand their mathematics and science education from a new and richer perspective.

The following section will discuss the TIMSS experience through the lens of the United States, addressing the various components of TIMSS and highlighting the major findings of the study as viewed by the United States. This discussion draws on publications that document the TIMSS achievement results as well as much of the background information both internationally and for the United States (see the Bibliography at the end of this chapter).

Mathematics		Science	
Country	**Mean Achievement**	**Country**	**Mean Achievement**
Singapore	625	Korea	597
Korea	611	Japan	574
Japan	597	▶ United States	565
Hong Kong	587	*Austria* [2]	565
Netherlands [2]	577	*Australia* [2]	562
Czech Republic	567	*Netherlands* [2]	557
Austria [2]	559	Czech Republic	557
Slovenia [3]	552	England	551
Ireland	550	Canada	549
Hungary [4]	548	Singapore	547
Australia [2]	546	*Slovenia* [3]	546
▶ United States	545	Ireland	539
Canada	532	Scotland	536
Israel [1,4]	531	Hong Kong	533
Latvia [1,2]	525	*Hungary* [4]	532
Scotland	520	New Zealand	531
England	513	Norway	530
Cyprus	502	*Latvia* [1,2]	512
Norway	502	*Israel* [1,4]	505
New Zealand	499	Iceland	505
Greece	492	Greece	497
Thailand [3,4]	490	Portugal	480
Portugal	475	Cyprus	475
Iceland	474	*Thailand* [3,4]	473
Iran, Islamic Republic	429	Iran, Islamic Republic	416
Kuwait [3,4]	400	*Kuwait* [3,4]	401
International Average	**529**	**International Average**	**524**

SOURCE: IEA Third International Mathematics and Science Study (TIMSS), 1994-95

☐ Significantly Higher than U.S. Average
☐ Not Significantly Different than U.S. Average
☐ Significantly Lower than U.S. Average

* Four years of formal schooling except in Australia and New Zealand – 4 or 5 years; England, Scotland, and Kuwait – 5 years; and Sweden – 3 years.

The Flemish and French educational systems in Belgium participated separately.

Countries shown in italics did not satisfy sampling guidelines in the following ways:

[1] Did not include all student populations. Israel tested students in the Hebrew Public Education System (72% of population) and Latvia tested students in Latvian-speaking schools (60% of population).

[2] Did not meet overall sample participation rate of 75% (product of school and student participation).

[3] Grades tested resulted in high percentage of older students.

[4] Unapproved classroom sampling procedures.

Figure 2.2. US average achievement in mathematics and science compared with other
TIMSS countries – fourth grade

Mathematics

Country	Mean Achievement
Singapore	**643**
Korea	**607**
Japan	**605**
Hong Kong	**588**
Belgium (Fl)	**565**
Czech Republic	**564**
Slovak Republic	**547**
Switzerland [1]	**545**
Netherlands [2]	*541*
Slovenia [3]	*541*
Bulgaria [2]	*540*
Austria [2]	*539*
France	**538**
Hungary	**537**
Russian Federation	**535**
Australia [2]	*530*
Ireland	**527**
Canada	**527**
Belgium (Fr) [2]	*526*
Sweden	**519**
Thailand [4]	*522*
Israel [1,4]	*522*
Germany [1,3]	*509*
New Zealand	**508**
England	**506**
Norway	**503**
Denmark [4]	*502*
▶ **United States**	**500**
Scotland [2]	*498*
Latvia [1]	**493**
Spain	**487**
Iceland	**487**
Greece [4]	*484*
Romania [3]	*482*
Lithuania [1]	**477**
Cyprus	**474**
Portugal	**454**
Iran, Islamic Rep.	**428**
Kuwait [4]	*392*
Colombia [3]	*385*
South Africa [4]	*354*
International Average	**513**

Science

Country	Mean Achievement
Singapore	**607**
Czech Republic	**574**
Japan	**571**
Korea	**565**
Bulgaria [2]	*565*
Netherlands [2]	*560*
Slovenia [3]	*560*
Austria [2]	*560*
Hungary	**554**
England	**552**
Belgium (Fl)	**550**
Australia [2]	*545*
Slovak Republic	**544**
Russian Federation	**538**
Ireland	**538**
Sweden	**535**
▶ **United States**	**534**
Germany [1,3]	*531*
Canada	**531**
Norway	**527**
New Zealand	**525**
Thailand [4]	*525*
Israel [1,4]	*524*
Hong Kong	**522**
Switzerland [1]	**522**
Scotland [2]	*517*
Spain	**517**
France	**498**
Greece [4]	*497*
Iceland	**494**
Romania [3]	*486*
Latvia [1]	**485**
Portugal	**480**
Denmark [4]	**478**
Lithuania [1]	**476**
Belgium (Fr) [2]	*471*
Iran, Islamic Rep.	**470**
Cyprus	**463**
Kuwait [4]	*430*
Colombia [3]	*411*
South Africa [4]	*326*
International Average	**516**

SOURCE: IEA Third International Mathematics and Science Study (TIMSS), 1994-95

☐ Significantly Higher than U.S. Average
☐ Not Significantly Different than U.S. Average
☐ Significantly Lower than U.S. Average

* Eight years of formal schooling except in Australia and New Zealand – 8 or 9 years; England, Scotland, and Kuwait – 9 years; Russian Federation and Sweden – 7 or 8 years; and Denmark, Norway, Sweden, and Philippines – 7 years.

The Flemish and French educational systems in Belgium participated separately.

The sample size was approximately 150 schools and 3,000 students in each country. Students were tested in both mathematics and science.

Countries shown in italics did not satisfy sampling guidelines in the following ways:

[1] Did not include all student populations. Israel tested students in the Hebrew Public Education System (74% of population); Latvia tested students in Latvian-speaking schools (51% of population); Lithuania tested students in Lithuanian speaking schools (84% of population); Germany tested 15 of 16 regions (88% of population); and Switzerland tested 22 of 26 cantons (86% of population).

[2] Did not meet overall sample participation rate of 75% (product of school and student participation).

[3] Grades tested resulted in high percentage of older students.

[4] Unapproved classroom sampling procedures.

Figure 2.3. US average achievement in mathematics and science compared with other
TIMSS countries – eighth grade

Mathematics

Country	Mean Achievement
Netherlands	560
Sweden	552
Denmark	547
Switzerland	540
Iceland	534
Norway	528
France	523
New Zealand	522
Australia	522
Canada	519
Austria	518
Slovenia	512
Germany	495
Hungary	483
Italy	476
Russian Federation	471
Lithuania	469
Czech Republic	466
▶ *United States*	**461**
Cyprus	446
South Africa	356
International Average	**500**

Science

Country	Mean Achievement
Sweden	559
Netherlands	558
Iceland	549
Norway	544
Canada	532
New Zealand	529
Australia	527
Switzerland	523
Austria	520
Slovenia	517
Denmark	509
Germany	497
Czech Republic	487
France	487
Russian Federation	481
▶ *United States*	**480**
Italy	475
Hungary	471
Lithuania	461
Cyprus	448
South Africa	349
International Average	**500**

SOURCE: IEA Third International Mathematics and Science Study (TIMSS), 1995-96

Significantly Higher than U.S. Average

Not Significantly Different than U.S. Average

Significantly Lower than U.S. Average

* Twelve years of formal schooling for most students in most countries, but there are a number of variations across and within countries with respect to the grades representing the final year of schooling for students.

The sample size was approximately 150 schools and 2,000 students in each country. Students were tested in both mathematics and science literacy.

Countries shown in italics did not satisfy sampling guidelines. With the exception of Germany, the italicized countries did not meet the overall sample participation rate of 75% (product of school and student participation). In addition, the following countries had unapproved classroom sampling procedures: the Netherlands, Denmark, Slovenia, Germany, and South Africa.

Figure 2 4. US average achievement in mathematics and science literacy compared with other TIMSS countries – the final year of secondary school

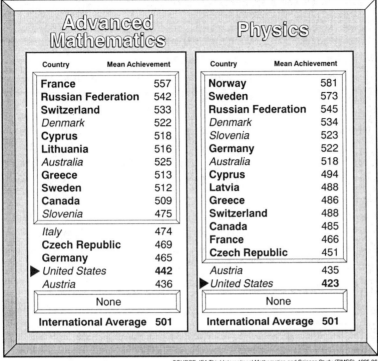

SOURCE: IEA Third International Mathematics and Science Study (TIMSS), 1995-96

Significantly Higher than U.S. Average

Not Significantly Different than U.S. Average

Significantly Lower than U.S. Average

* Twelve years of formal schooling for most students in most countries, but there are a number of variations across and within countries with respect to the grades representing the final year of schooling for students.

The sample size was approximately 100 schools and 1,000 students in each country for each test.

The countries shown in italics did not meet overall sample participation rate of 75% (product of school and student participation). In addition, Denmark and Slovenia had unapproved classroom sampling procedures.

Figure 2.5. US average achievement in advanced mathematics and physics compared with other TIMSS countries – the final year of secondary school

3.1 Mathematics and science achievement

3.1.1 Fourth Grade

US fourth-grade students performed well on TIMSS, especially in science. Figure 2.2 shows the performance of US fourth graders in mathematics and science compared to that of similar-aged students in the 25 other countries participating in this part of the achievement testing. In mathematics, fourth-grade students in seven countries significantly outperformed US fourth graders, and in six additional countries performance was not significantly different from that of US students. US fourth graders outperformed their counterparts in 12 nations.

In science, students in only one country, Korea, significantly outperformed US fourth graders. Student performance in science in five countries (Japan, Austria, Australia, the Netherlands, and the Czech Republic) was not significantly different from US fourth graders. US fourth graders outperformed their counterparts in science in 19 nations. Looked at from a different perspective, US fourth graders performed above the international average of the 26 TIMSS countries in both mathematics and science.

In mathematics, Singapore and Korea were the top-performing countries. Japan and Hong Kong also performed among the best in the world, as did the Netherlands, the Czech Republic, and Austria. In science, Korea was the top-performing country. In addition to the United States, Japan, Austria, and Australia also performed very well in science.

3.1.2   Eighth Grade

The relative standing of US eighth graders was weaker than that of fourth graders in both mathematics and science. Figure 2.3 presents the TIMSS achievement results for the 41 countries participating at the eighth grade. In mathematics, students in 20 countries significantly outperformed US eighth graders. Mathematics achievement for eighth-grade students in 13 countries was not statistically significantly different from that of US students. US students outperformed their counterparts in 7 countries in mathematics.

Compared to their international counterparts, US students were below the international average in mathematics and above the international average in science. In science, students in nine countries significantly outperformed US eighth graders. Performance in 16 other countries was not significantly different than that of US students. The US students significantly outperformed 15 countries in science. Singapore was the top-performing country in both mathematics and science. Korea, Japan, and the Czech Republic also performed very well in both subjects as did Hong Kong and the Flemish part of Belgium in mathematics. In the mathematics and science content areas, US eighth graders performed below the international averages for physics and chemistry.

3.1.3   Twelfth Grade

Given the extensive diversity of students' curricula in the upper secondary school, there is wide variation in student preparation in mathematics and science, and the assessment of student knowledge in these subjects consequently poses a particular challenge. In response to this challenge, TIMSS developed three different tests. The mathematics and science literacy test was designed to measure application of general mathematics and science knowledge for all final-year students regardless of their school curriculum. For example, the tests included questions related to the environment, health, home maintenance (e.g., how much paint is necessary to paint a room, how many tiles will cover a floor, and how much lumber is needed to build a bookcase), consumer issues related to comparison shopping, and interpretation of graphs such as those often presented in newspapers or on television. Since the purpose of this part of the testing was to see how well students could use what they had learned in situations likely to occur in their everyday lives, it represented mathematics and science content covered in most countries no later than about the ninth grade. Twenty-one countries participated in this part of the testing, which also was designed to provide results separately for mathematics and science.

There also was great interest on the part of some TIMSS countries to determine what school-leaving students with special preparation in mathematics and science know and can do, since the capabilities of these students may help determine a country's future potential to compete in a global economy. Thus, a second test was developed for students having taken advanced mathematics. For science, it was not possible to study all branches of science in detail. The participating countries chose physics for detailed study because it is the branch of science most closely associated with mathematics, and comes closest to embodying the essential elements of natural science. The third test, then, was a physics test designed to measure learning of physics concepts and knowledge among final-year students having studied physics.

In mathematics literacy, US twelfth-grade students performed below the international average, and their achievement was among the lowest of the 21 countries (see Figure 2.4). Students in the final year of secondary school in 14 countries scored significantly above US twelfth graders. Students in four countries did not score significantly different from the United States and students in only two countries performed below the United States (Cyprus and South Africa).

Consistent with patterns at grades four and eight, the US students' relative performance in science literacy was slightly better than in mathematics literacy (shown in Figure 2.4). Still, US students performed below the international average, and their achievement in science literacy was among the lowest of the 21 countries. US twelfth graders were significantly outperformed in science literacy by final-year students in eleven countries. They performed similarly to final-year students in seven countries, and were better than students in only two countries (Cyprus and South Africa).

The results for advanced mathematics and for physics are presented in Figure 2.5. The US twelfth-grade students did not significantly outperform the final-year students in any of the other participating countries in either advanced mathematics or physics. Of the 16 participating countries[2.4], the US ranked second to last in advanced mathematics and last in physics.

3.1.4   Contexts for Achievement Results

TIMSS enabled the United States to go beyond obtaining a status report on achievement in mathematics and science to providing possible explanations for the achievement results. Through its many components, TIMSS provided information about the contexts for learning mathematics and science in the United States and how this compares with practices in other countries.

3.1.5   Curriculum

Through the teacher and student questionnaires, and the TIMSS curriculum analysis, the United States learned from TIMSS that the content and structure of the US mathematics and science curricula are less rigorous that those in high-performing countries. In mathematics, students in the United States experience more tracking and repetition of topics already taught, rather than moving ahead to more challenging material. Fewer eighth graders study algebra and geometry than in other countries, and fewer twelfth graders are taking mathematics (only 66%) than in other countries. In science, US eighth graders are relatively weak in physics and chemistry. In the United States, students study the sciences in a sequence, one per year, while in other countries, students tend to study more than one science simultaneously, for multiple years. As with mathematics, fewer twelfth graders are taking science than in other countries (47%).

[2.4] Sixteen countries participated in each component, although they are not quite the same countries. Lithuania and Italy have results only for advanced mathematics, whereas Norway participated only in the physics testing. Note that Greece participated in the two components of the specialised testing, even though it did not have the resources to test a sample of all students in literacy. Similarly, Latvia participated in only the physics testing.

3.1.6   Quality of Instruction

Through the TIMSS video study of mathematics classrooms, the United States obtained some evidence that the quality of classroom mathematics instruction is low compared with that in Japan and Germany, the two other countries that participated in the video study. The videotaped lessons were evaluated for the coherence of sequencing; cognitive complexity; and the way problems and activities contributed to developing mathematical concepts. The pedagogical approach used by US eighth-grade mathematics teachers revolved around the teacher explaining a concept to students and then students working on their own. And, as shown in Figure 2.6, US teachers rarely developed concepts for students, in contrast to teachers in Germany and Japan.

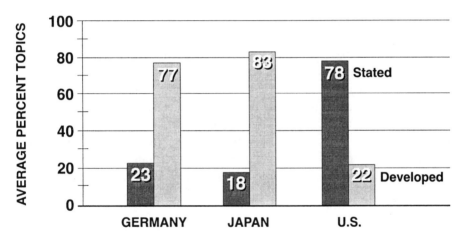

Figure 2.6. Average percentage of topics in eight-grade mathematics lessons that are stated or developed

Source: TIMSS; *Pursuing Excellence*, NCES 97-198 Videotape Classroom Study, UCLA, 1995

The TIMSS results also show that US teachers do not practice the recommendations of the mathematics reform movement in terms of pedagogy, although they think that they do.

3.1.7   Time on Task

The United States learned from the videotape study that US eighth-graders may spend less time studying mathematics and science than students in other countries. More in-class instructional time is spent on homework (at the eighth grade) and US lessons were fraught with interruptions and discussions of topics unrelated to the learning of mathematics. The student questionnaires showed that US twelfth graders spend more time working at a paid job than final-year students in other countries (three hours per day, on average).

3.1.8   Teacher Preparation and Support

Through the case studies carried out in the United States, Germany, and Japan, the United States learned that US beginning teachers lack the apprenticeship and systematic mentoring afforded to teachers in Germany and Japan. The United States also has learned that it has many 'out-of-field' teachers. Fifty-five per cent of high school physical science teachers do not have even a minor in any of the physical sciences.

3.1.9   Motivation for Students

Finally, it is possible that motivation for students to achieve is not as high in the United States as in other countries. For better or worse, the United States does not have a formal examination system, which, of course, can be a motivating factor. Compared to the high-performing Asian countries, there is less parental and cultural support for learning in the United States. Finally, there are many more unjustified absences from school in the United States compared to the high-performing countries, pointing to students' lack of motivation and complacency on the part of parents and schools.

3.1.10   Reassuring Findings

Not all of the TIMSS news for the United States was bad news. The notion that US students spend far more time in front of the television than students in other countries was dispelled; TIMSS found that students in other countries also spend considerable amount of time watching television and talking with friends. The United States found fewer achievement differences between boys and girls than were found in many of the participating countries. Finally, US students reported very positive attitudes towards mathematics and science.

3.1.11   TIMSS-Repeat and Benchmarking for States and Districts

The success of TIMSS as a source of information for educators, policy makers, and the public, led the United States government to explore the possibility of conducting TIMSS again to measure trends in achievement from 1995 to 1999. TIMSS-R has been conducted at the eighth-grade in

1999 in about 40 countries. Funding for the international co-ordination of TIMSS-R is provided by the US National Centre for Education Statistics, the US National Science Foundation, and the World Bank. In addition, the video study is being carried out again in the United States and about ten other countries, and will include observations of mathematics and science classrooms. The video study is being partially funded by the National Centre for Education Statistics.

TIMSS-R is designed to provide trend data for those countries that participated in TIMSS in 1995. Countries that participated in 1995 can see how their eighth-grade students are performing four years later. Countries that tested students in the upper grade of Population 1 in 1995 can see how that cohort is performing as eighth graders. And, countries that did not participate in 1995 have joined TIMSS-R to obtain international mathematics and science comparisons for the first time.

The TIMSS-R tests and questionnaires were developed, field-tested, and prepared for administration in 1998/9. The test development involved replacing the TIMSS 1995 achievement items that were released to the public. The TIMSS 1995 questionnaires were shortened, where appropriate, and updated. Data collection began in October 1998 in countries on a Southern Hemisphere schedule and in March 1999 in countries on a Northern Hemisphere schedule.

Another US initiative that grew out of the interest surrounding the TIMSS 1995 results and TIMSS-R, is the Benchmarking for States and Districts, funded by the National Centre for Education Statistics and the National Science Foundation. TIMSS Benchmarking provided states and districts with the opportunity to administer the TIMSS-R tests in the spring of 1999, at the same time as students were tested around the world, and to compare their students' achievement with that of students in the TIMSS-R countries. By September 1998, 25 states and districts had enrolled in this program.

BIBLIOGRAPHY

TIMSS International Reports
Beaton, A. E., Martin, M. O., Mullis, I. V. S., Gonzalez, E. J., Smith, T. A., and Kelly, D. L.
(1996a) *Science Achievement in the Middle School Years: IEA's Third International Mathematics and Science Study (TIMSS)* (Chestnut Hill, MA: Boston College)
Beaton, A. E., Mullis, I. V. S., Martin, M. O., Gonzalez, E. J., Kelly, D. L., Smith, T. A.
(1996b) *Mathematics Achievement in the Middle School Years: IEA's Third International Mathematics and Science Study (TIMSS)* (Chestnut Hill, MA: Boston College)
Harmon, M., Smith, T. A., Martin, M. O., Kelly, D. L., Beaton, A. E., Mullis, I. V. S.,
Gonzalez, E .J., and Orpwood, G. (1997) *Performance Assessment in IEA's Third International Mathematics and Science Study* (Chestnut Hill, MA: Boston College)

Martin, M. O., Mullis, I. V. S., Beaton, A. E., Gonzalez, E. J., Smith, T. A., and Kelly, D. L. (1997) *Science Achievement in the Primary School Years: IEA's Third International Mathematics and Science Study (TIMSS)* (Chestnut Hill, MA: Boston College)

Mullis, I. V .S., Martin, M. O., Beaton, A. E., Gonzalez, E. J., Kelly, D. L., and Smith, T. A. (1997) *Mathematics Achievement in the Primary School Years: IEA's Third International Mathematics and Science Study (TIMSS)* (Chestnut Hill, MA: Boston College)

Mullis, I. V. S., Martin, M. O., Beaton, A. E., Gonzalez, E. J., Kelly, D. L., and Smith, T. A. (1998) *Mathematics and Science Achievement in the Final Year of Secondary School: IEA's Third International Mathematics and Science Study (TIMSS)* (Chestnut Hill, MA: Boston College)

Technical Reports

Martin, M. O. and Kelly, D. L., (eds.) (1996) *TIMSS Technical Report, Volume I: Design and Development* (Chestnut Hill, MA: Boston College)

Martin, M. O. and Mullis, I. V. S. (eds.) (1996) *Quality Assurance in Data Collection* (Chestnut Hill, MA: Boston College)

Martin, M. O. and Kelly, D. L., (eds.) (1997) *TIMSS Technical Report, Volume II: Implementation and Analysis, Primary and Middle School Years* (Chestnut Hill, MA: Boston College)

Martin, M. O. and Kelly, D. L., (eds.) *TIMSS Technical Report, Volume III: Implementation and Analysis, Final Year of Secondary School* (Chestnut Hill, MA: Boston College)

Released Item Sets

TIMSS Mathematics Items, Released Set for Population 1 (Primary) (Chestnut Hill, MA: Boston College)

TIMSS Science Items, Released Set for Population 1 (Primary) (Chestnut Hill, MA: Boston College)

TIMSS Mathematics Items, Released Set for Population 2 (Middle) (Chestnut Hill, MA: Boston College)

TIMSS Science Items, Released Set for Population 2 (Middle) (Chestnut Hill, MA: Boston College)

TIMSS Released Item Set for the Final Year of Secondary School: Mathematics and Science Literacy, Advanced Mathematics, and Physics (Chestnut Hill, MA: Boston College)

TIMSS Encyclopaedia

Robitaille, D. F. (ed.) (1997) *National Contexts for Mathematics and Science Education: An Encyclopaedia of the Education Systems Participating in TIMSS* (Vancouver: Pacific Educational Press)

Curriculum Analysis Reports

Schmidt, W. H., McKnight, C. C., Valverde, G. A., Houang, R. T., and Wiley, D. E. (1997a) *Many Visions, Many Aims: A Cross-National Investigation of Curricular Intentions of Curricular Intentions in School Mathematics* (Dordrecht: Kluwer)

Schmidt, W. H., Raizen S. A., Britton, E. D., Bianchi, L .J., and Wolfe, R. G. (1997b) *Many Visions, Many Aims: A Cross-National Investigation of Curricular Intentions in School Science* (Dordrecht: Kluwer)

US TIMSS Reports

Stevenson, H. W. and LeTendre, G. (eds.) (1998) *The Educational System in Japan: Case Study Findings*. Publication No. SAI98-3008 (Washington, DC: US Department of Education, National Institute on Student Achievement, Curriculum, and Assessment)

Stevenson, H. W. and LeTendre, G. (eds.) *The Educational System in Germany: Case Study Findings* (Washington, DC:,US Department of Education, National Institute on Student Achievement, Curriculum, and Assessment)

Stevenson, H. W. and LeTendre, G. (eds.) *The Educational System in the United States: Case Study Findings* (Washington, DC: US Department of Education, National Institute on Student Achievement, Curriculum, and Assessment)

United States Department of Education, National Centre for Education Statistics (1996) *Pursuing Excellence: A Study of US Eighth-Grade Mathematics and Science Teaching, Learning, Curriculum, and Achievement in International* Context, NCES pp. 97-198 (Washington, DC: US Government Printing Office)

United States Department of Education, National Centre for Education Statistics (1997) *Pursuing Excellence: A Study of US Fourth-Grade Mathematics and Science Achievement in International Context*, NCES pp. 97-255 (Washington, DC: US Government Printing Office)

United States Department of Education, National Centre for Education Statistics (1998) *Pursuing Excellence: A Study of US Twelfth-Grade Mathematics and Science Achievement in International Context*, NCES, pp. 98-049 (Washington, DC: US Government Printing Office)

United States Department of Education, Office of Educational Research and Improvement (1997) *Moderator's Guide to Eighth-grade Mathematics Lessons: United States, Japan, and Germany* (*Washington*, DC: US Government Printing Office)

Further information regarding TIMSS (including various publications) can be found on the following web sites: www.csteep.bc.edu/timss www.ed.gov/NCES/timss

Chapter 3

Diversity of Purpose in International Assessments
Issues arising from the TIMSS Tests of Mathematics
and Science Literacy

Graham Orpwood
York University, Canada

TIMSS was the largest international assessment project ever undertaken, however such things can be measured. It began in 1990 and is still in progress. It has cost many millions of dollars to conduct and involved more researchers, more schools and more students in more countries than any previous comparative study. Over 40 countries participated in the core study of 13 year-olds with more electing to join for the TIMSS-R replication study in 1999.

Of course not only researchers in each country were involved. TIMSS has involved policy-makers in Ministries of Education, funding agencies, subject matter and assessment experts, school administrators, and teachers. This involved many thousands of people across all the participating countries. In addition the reports in each country were devoured by the media and read by parents and many others who have an interest in science and mathematics in schools. As Science Co-ordinator for the first part of TIMSS, I was heavily involved in the test development component of the study and so what follows is very much a personal perspective and in no way reflects an official TIMSS position.

We can never know what all of these people who gave their time and professional energies to TIMSS did it for, or even what everyone expected to come out of such a major international study. However, we can get a glimpse of some of the purposes of TIMSS from the written reports and background papers by people involved in the study and this question of purpose of international assessment is the issue I would like to reflect on briefly here. I will use these reflections as a context for discussing aspects of the Mathematics and Science Literacy (MSL) assessment which was an

D. Shorrocks-Taylor and E.W. Jenkins (eds.), Learning from Others, 49–62.
© 2000 *Kluwer Academic Publishers. Printed in the Netherlands.*

element of the third part of TIMSS focusing on students in their final year of high school.

Michael Martin and Ina Mullis, Deputy Study Directors of TIMSS, have referred in Chapter 2 to the study's vision, its challenges, and its final reality and of course all such studies have these characteristics. They acknowledge that the challenges inherent in undertaking TIMSS were so formidable that fully achieving the original vision in reality was impossible and it would be unreasonable to disagree with that sentiment. Further, they have argued that the reality of TIMSS as it was achieved has been an important and valuable contribution both to our understanding of cross-national differences in mathematics and science and also to policy-making in many of the participating countries. I would endorse this view.

Nonetheless, it is also useful, in my view, to examine the vision, the challenges and the reality more closely and critically to see what compromises were made along the way and to try to learn from the TIMSS experience something about how such international studies can be improved in the future. In raising such questions, I am not – as some critics have tried to do – attempting to undermine the value of conducting either this study or others like it. Rather I am aiming to derive lessons for the future while the memory of the recent past is still fresh.

1.   WHAT WAS THE PURPOSE OF TIMSS?

Reviewing some of the published documents related to TIMSS reveals a variety of perceived purposes or visions of the study. Consider the following statement, for example:

> Broadly stated, the goal of TIMSS is to learn more about mathematics and science curricula and teaching practices associated with high levels of student achievement, in order to improve the teaching and the learning of mathematics and science around the world. (Robitaille and Robeck, 1996)

David Robitaille, the principal author of this statement, was the initiator of TIMSS and for the first three years (1990-93) its International Coordinator. His background is as a mathematics teacher, textbook author, and teacher educator in Canada. He was also heavily involved in the Second International Mathematics Study (SIMS) in the 1980s. It is, in my view, no coincidence therefore that the emphasis in his statement of the goal of TIMSS is on curriculum, on teaching, and on the improvement of practice.

Compare this with a second statement of the purpose of TIMSS:

Its (TIMSS's) aims are not only to compare the achievements of students in many different countries but to explain how countries differ in what they teach, the way they teach, and in how they overcome the obstacles to student learning. (Beaton 1996)

Al Beaton of Boston College took over from David Robitaille as International Study Director of TIMSS in 1993. He brought to TIMSS the experience of over 20 years at the (US) Educational Testing Service (ETS) with particular responsibility for technical, statistical and methodological features of the National Assessment of Educational Progress (NAEP). His interest and expertise is in comparative assessment of achievement and his statement of the aims of TIMSS, while not inconsistent with Robitaille's, nevertheless reflects a different emphasis.

Finally, consider yet a third statement about the purpose of TIMSS:

TIMSS is a study of education systems and how they define, deliver, and organise instruction to provide educational opportunities to groups of students. (Schmidt and McKnight, 1998)

The principal author of this statement, Bill Schmidt, an educational researcher with a psychometric background from Michigan State University, was the co-ordinator of the curriculum analysis components of TIMSS, one of the team that developed the original conceptualisation for TIMSS, and the National Research Co-ordinator for TIMSS in the United States. Once again the emphasis of this statement, while not inconsistent with the other two, is quite distinct.

All the statements of purpose cited here are from people centrally involved with the international planning and development of TIMSS. In addition to these views, it would be interesting to know what were the various purposes for which each participating country joined TIMSS, and whether these too would have varied depending on whether one asked the National Research Co-ordinator (NRC), the funding agency/Ministry of Education (where these were different), or the mathematics and science teachers' organisations. Certainly, in the case of the United States – where, at a 1996 conference in San Francisco, the question was put to a panel discussion – the TIMSS study director, Al Beaton, repeated essentially what is attributed to him here while another panellist, Mary Lindquist, then the President of the (US) National Council of Teachers of Mathematics, claimed that the purpose of TIMSS was 'the improvement of mathematics teaching.' Personally, I would not be surprised to find the equivalent diversity in many of the countries that participated in TIMSS.

So there is a variety of purposes for which people and organisations become involved in TIMSS. But, we may ask, does it make a difference? After all, there are many co-operative activities in life, in which the participants have different reasons for participating, but which nonetheless function effectively and to the equal satisfaction of all. In the case of international assessment, of which TIMSS is currently the prime example, I will argue here that it can and does make a difference, both to how several key aspects of the project were carried out and to the subsequent satisfaction that people have with the outcome.

2.   MATHEMATICS AND SCIENCE LITERACY IN TIMSS

In this chapter, I will use, as the vehicle for my argument, the study of Mathematics and Science Literacy (MSL) which was undertaken as part of the third component of TIMSS, since it is an aspect of the study in which I was personally involved.[3.1] This component of TIMSS involved a population of students in their final year of secondary school in some 22 countries[3.2]. In the words of the final international report, the MSL test was designed 'to provide information about how prepared the overall population in each country is to apply knowledge in mathematics and science to meet the challenges of life beyond school' (Mullis *et al.* 1998, p. 13).

Unlike both other components of TIMSS and other IEA studies, the MSL study was not directly curriculum-based. That is, it was not an attempt to measure what had been taught and learned in a given year of schooling or to a given age-group of students. Instead, it was a study of the mathematics and science learning that final-year students have retained regardless of their current areas of study. These students may have studied mathematics and science in their final years of school or they may not have; they may regard themselves as specialists in mathematics and science, in other subjects, or in none; they may be entering occupations or further education related to mathematics and science, or they may have no intention of doing so. Nonetheless, all of them have studied mathematics and science at some time during their school careers and all of them are entering a world increasingly affected by science and technology. The role of the MSL study within TIMSS, therefore, was to ask whether school leavers could remember the

---

[3.1] Further details of the MSL component of TIMSS can be found in Orpwood and Garden (1998), in Mullis, Martin, Beaton, Gonzalez, Kelly and Smith (1998) and in the national reports of participating countries.

[3.2] For details of how this was defined operationally, see Mullis *et al.* (1998) Appendix B.

mathematics and science they have been taught and could apply this knowledge to the challenges of life beyond school.

In this paper, I shall review two aspects of the MSL study – the test development and the data analysis and reporting – and use these to reflect on how the diverse purposes for the overall study influenced these activities and their results.

3.   MSL TEST DEVELOPMENT

It was one thing to decide to include a test of mathematics and science literacy in TIMSS. It was quite another to translate that decision into a conceptualisation of MSL and from there to assemble test items that were a reasonably valid measure of the concept. Undertaking the task would have been daunting in any circumstances, but in a study involving as many countries as TIMSS, it presented particular difficulties. The task involved two processes. More exactly, it was a single task of two dimensions, each with the same end in view but each also having its own context, purpose, criteria for success, and sense of time.

The first dimension comprised a conceptual inquiry leading to a precise, defensible definition of mathematics and science literacy. The inquiry was set in the context of ongoing debates in the scholarly and professional literature about appropriate goals for mathematics and science education, about the needs of young people leaving school and entering a complex world dominated by technology, and about the nature of mathematics, science, and technology themselves. This dimension, defining the concept, was essentially theoretical and scholarly, and represented an area in which leaders of mathematics, science, and technology education have been engaged for over three decades. The definition of MSL was undertaken by subject specialists removed from day-to-day involvement with TIMSS. It had its own debates and positions, its own advocates and defenders, and its own criteria for success. The time lines were assumed to be open-ended, as is always the case with scholarly inquiry, creating a conflict with the practical dimension of creating a literacy test. The criteria underlying the enquiry were based on those of rational argument and analysis, since the literacy concept had to be defensible in the courts of scholarly criticism.

The second dimension of the task was the development of test instruments for use in the MSL component of TIMSS. This was essentially a practical task, undertaken by subject-matter specialists close to the TIMSS community and having definite time lines, although these were extended twice during the course of the process. The test instrument development had political as well as conceptual criteria for success, having not only to be based on a defensible level of construct validity but also to be acceptable to the countries participating in the MSL component of TIMSS. The process was therefore less one of scholarly debate than an extended negotiation relating to the literacy concept itself, the items proposed for use in the tests, and the constraints inherent in the test design and other aspects of TIMSS.

The original plan was first to define and clarify the concept of MSL and then to develop appropriate test items. As with all such two-dimensional conceptual and practical tasks, however, the two components could not take place sequentially. The process took nearly four years, with the conceptual dimension occupying much of the early stages and the practical one, much of the latter. From the outset, however, potential test items were being reviewed by groups of TIMSS participants and to the end, the MSL concept was being debated and adapted, some might say compromised, to ensure that the literacy test could actually take place. The two dimensions were thus continuously linked throughout the four-year period.

What resulted from this intriguing blend of political and conceptual inquiry was a framework for MSL, which comprised four major elements:

– *Mathematics Content*
  Basic number operations; tables and graphs; spatial relations; algebraic sense; estimation; proportionality; measurement; likelihood of events; sampling and inferences
– *Science Content*
  Human biology; earth and space science; energy; physical science
– *Reasoning in Mathematics, Science, and Technology*
  Informal reasoning; use of data and display; translation between graphical/quantitative information including statistical to-and-from natural language statements; problem identification; mathematical or scientific ways of knowing
– *Social Impacts of Mathematics, Science and Technology*
  Issues in science, technology and environment (pollution, conservation of resources, agriculture, energy-related technology, population/health, information technology, biotechnology); social impacts of mathematics and statistics (polls, surveys, cost of living and economic indicators, gambling and insurance, measuring risk/making decisions in the face of uncertainty, trends and predictions, quantitative input to a larger information package, quality control and performance measurement, deciding on the bases of planned experiments).

The third and fourth categories of the framework were subsequently combined into one called Reasoning and Social Utility in mathematics, science, and technology (RSU). The final framework, including a blueprint for the MSL tests, was summarised in a short paper which was discussed and approved by the TIMSS Subject Matter Advisory Committee and by TIMSS

National Research Co-ordinators at their November 1993 meeting in Frascati, Italy. It was subsequently used as the basis for development of the MSL tests.

Meanwhile, the collection, development and field-testing of potential items for the MSL tests were taking place under the leadership of staff at the Australian Council for Educational Research (ACER). The final selection of MSL items was also constrained by a number of factors that severely limited the scope of what could be assessed. These included the conceptual framework described earlier but also the following:

- Consistency with the conceptual framework described earlier; for example, items selected for the literacy test should, wherever possible, be contextualised in real life situations.
- The need to include a balance of earth science, life science, and physical science within the science content area but also to include a number of 'link items' – items appearing in both the MSL test and in the test of 13 year-olds (TIMSS Population 2) – and a sufficient number of items corresponding to two 'in-depth topics': Human Biology and Energy Types, Sources, and Conversions[3.3]. A corresponding balance was sought in mathematics.
- Maintaining a balance among item types: multiple-choice (MC), short answer (SA) and extended response (ER) items.
- The total time available for literacy testing in the Population 3 test design (see Adams and Gonzalez, 1996): four 30-minute clusters of items were required. These included one cluster of items representing Reasoning and Social Utility (RSU) and three clusters each containing equal amounts of time devoted to mathematics and science. This meant that science content was restricted to 45 minutes of testing time.
- The use of items whose psychometric properties and level of acceptability to participating countries (based on the field trials) fell within an acceptable range[3.4].

These factors meant that the overall literacy item pool in the MSL survey totalled just 76 items as shown in Table 3.1.

Table 3.1. Composition of MSL Item Pool

Scale	Item Type			Total Number of Items	Total Testing Time (mins)
	MC	SA	ER		
Mathematics Content	31	7	0	38	45
Science Content	16	7	3	26	45
RSU	5	3	4	12	31
TOTAL	52	17	7	76	121

[3.3] The special focus on certain 'in-depth' topics derives from the original TIMSS design and particularly from that of the curriculum analysis (see Schmidt *et al.*, 1997).

[3.4] For more details of how this was achieved, see Garden and Orpwood (1996).

On reflection, I believe that the selected MSL items reflect a greater adherence to what can be called the four 'technical' criteria as detailed above than to the spirit of the conceptual framework referred to above. For example, many draft items that went beyond strict knowledge of science or mathematics content were either eliminated on psychometric grounds or on the grounds of unacceptability to participating countries. This affected the selection of RSU items in particular and, with a few exceptions, the items used represented the 'lowest common denominator' of acceptability, rather than a ground-breaking novel assessment.

This is of particular concern to those countries in which the recent focus of science educators on Science, Technology and Society (STS) is designed to increase students' ability to relate their scientific knowledge to the world beyond the school. In such countries, items that are aimed at assessing students' achievement of these newer goals of the curriculum are very thinly represented in TIMSS. In such a situation, the desire for high validity in the MSL assessment can be seen to have given way to the desire for high reliability, a tendency noted by many critics and commentators on assessment in recent years[3.5]

This is not to say that all the concessions were made by the subject-matter specialists in favour of the psychometric considerations. Part of the problem derived from the difficulty and lack of experience in measuring some of the newer goals of science education such as those associated with STS. Part also derived from the fact that the very notion of mathematics and science 'literacy' is one that has had much greater debate within the English-speaking literature and that items that assessed students understanding of the impact of science and technology in a social or economic context were not well received among subject specialists in some TIMSS countries. As TIMSS Science Co-ordinator, I recognise that the pressure of time, especially toward the end of the test development, resulted in the exclusion of items that given more time might have increased the validity of the MSL test. Nonetheless, the resulting test seems to reflect a less than ideal resolution of the tensions inherent in the criteria we used.

4.   ANALYSIS AND REPORTING OF MSL ACHIEVEMENT

When the tests had been written and the item data cleaned and analysed, the results were scaled using item-response theory (IRT) to summarise the

---

[3.5] For example, Messick, Wiggins, Black.

achievement of participating countries[3.6]. For the MSL test, two scales were developed, called 'mathematics literacy' and 'science literacy', each with a mean of 500 and a standard deviation of 100. Each scale included all those items in one content area plus the corresponding RSU items. Thus the 'mathematics literacy' scale comprised the mathematics content items together with the most mathematics-related RSU items; the same was the case for the 'science literacy' scale. The composite 'MSL' results were computed by averaging the results on mathematics and science literacy scales. The international reports thus contained cross-national comparisons of these three results together with correlations with selected background variables, such as gender.

The results of the MSL assessment (along with the advanced mathematics and physics assessments in the same population) presented the Study Centre with many other more complex problems also. Because of the differing structure of school systems and the difficulties that many countries encountered in meeting the TIMSS sampling requirements, only eight of the 22 participating countries are presented in the international reports as having fully met the sampling guidelines. Furthermore, even in these eight countries, the proportion of school-levers covered by the TIMSS tests varied from a low of 43 per cent to a high of 82 per cent[3.7]. These factors make cross-country comparisons of MSL extremely complex.

In the case of Canada, for instance, the MSL achievement tables show that three countries significantly outperformed Canada, seven showed achievement levels not significantly different from those of Canada, and the remaining ten countries had results significantly lower than Canada's. With the exception of one country (South Africa) however, all the countries' results were bunched quite closely together, spanning little over 1 standard deviation (100 scale points) from top (Netherlands at 559) to bottom (Cyprus at 447). Little correlation was found between achievement and 'coverage.' In fact, countries with higher coverage showed slightly better MSL results overall (correlation of 0.56) which is perhaps counter to the popular assumptions.

5.   WHAT CAN BE LEARNED FROM THE TIMSS MSL STUDY?

In evaluating the outcomes of the MSL study it is instructive to go back to the three visions of the purpose of TIMSS articulated earlier. In short, one

---

[3.6] For more details see Mullis *et al.* (1998) Appendix B.
[3.7] For more on the TIMSS Coverage Index (TCI), see Mullis *et al.* (1998) pp 17-20.

or more of these three purposes motivated the many participating individuals and groups.

1. Learning more about curricula and teaching associated with high achievement in order to improve teaching and learning;
2. Comparing achievements and explaining how countries differ in teaching approaches;
3. Studying how educational systems differ in how they define, deliver and organise instruction.

The first of these purposes underpinned much of the early work in the development of the MSL study as the original focus was on the broad goals of science and mathematics education in school, particularly as these applied to the whole population (rather than to mathematics and science specialists). The hope was that we could see through the MSL study which aspects of science and mathematics were retained by students and which were not, and if students could apply what they knew to real world contexts. In particular, the one component of the MSL conceptual framework that was distinct from that of the tests of 13 year-olds (TIMSS Population 2) and 9 year-olds (TIMSS Population 1) was the RSU component. We were interested in seeing the extent to which students were able to reason about issues in which mathematics, science and technology impacted on society and whether this ability differed from simple knowledge of mathematics and science.

The results, as presented in the international reports, make it hard to see how this purpose can be achieved, at least from the first sets of international and national reports.[3.8] From a teacher's perspective, it is hard to see that the results of the MSL study presented in the international reports suggest anything that might influence decisions on either curriculum, teaching or even school improvement. Teachers, parents, and those whose primary professional commitment is to the improvement of teaching and learning, have every reason, it seems to me, to be disappointed in the outcomes of the TIMSS MSL study, regardless of where their country is placed in the league tables. Certainly, secondary analyses of TIMSS data can be made (see, for example, Orpwood, 1998 and the work of the Norwegian researchers in Chapter 8 in this volume) and implications for improving practice derived, but without tests that truly measure what they were intended to measure, this secondary analysis is severely limited in its scope.

The second purpose is clearly the one that operationally dominated both the completion of the test development and also the data analysis and reporting of TIMSS. To develop reliable comparative statistics, the IRT scaling methodology demanded that items be as discriminating as possible and with p-values within a medium band (i.e. none too easy or too difficult).

[3.8] An attempt has been made to analyse the Population 2 results in Ontario in more depth to discover what can be learned from TIMSS (Orpwood 1998) with limited success.

When item statistics from the field trial were not satisfactory, items were dropped from the test. When, in the final tests, the RSU cluster of items were not seen to have significantly differing item statistics, the category was simply dropped and the items added to the mathematics and science content scales. No attempt has yet been made to compare the achievement of science and mathematics specialists with the rest of the population on RSU despite the fact that these items had been built into the one cluster that appeared in all test booklets for this precise purpose.

Nonetheless, cross-national comparative tables ('league tables') were generated and some correlational studies conducted. Those whose interest is focussed mainly on such matters, including the media, politicians, and the critics of education, were undoubtedly satisfied. In my own jurisdiction (Ontario), the appearance of the first TIMSS report had a significant and immediate impact on the government, which promptly (within a week of the report's publication) contracted the acquisition of a new curriculum in science for Grades 1-8. Curiously, this form of presentation also had some unintended negative effects for some countries, as this anecdote following the publication of the Population 2 results illustrates. On being congratulated (by the writer) on the high achievement in her country – incidentally, a country with a rapidly developing industrial economy in which mathematics and science skills are of considerable economic importance – it was observed that such a high ranking might now make it *more* difficult to obtain financial support from the government for improving mathematics and science in her country.

The third purpose for TIMSS is arguably the most ambitious and interesting of all, at least from a research perspective. According to Schmidt and McKnight (and many others involved in TIMSS in participating countries would agree with them), 'it (TIMSS) was hardly conceived as primarily an exercise in cross-national rankings.' They continue:

> TIMSS was undergirded by a conceptual model based on the notion that educational systems are organised to provide students with access to potential learning experiences... The model describes various factors associated with schooling that define and influence the quality of those educational opportunities. The major organising questions are: What are the goals and intentions of an educational system? Who delivers the instruction? How is instruction organised? What have students learned? Only the last question potentially focuses on test score rankings. (Schmidt and McKnight, 1998, p. 1)

The ambitious research goals implied by the conceptual model referred to here may have always been beyond the reach of any single project. Certainly they have yet to be achieved in the TIMSS reports published to date. And it appears that once the 'league tables' are published, the level of political interest and the research funding that flows from that may be in decline.

In the case of MSL, it would be of great interest to understand more about the provisions made in different countries for students to be prepared to use science and mathematics in the world they face after school, how teachers are themselves prepared for such teaching, and the extent to which students achieve the real-life skills that are science- and mathematics-related. What made TIMSS unique was not just its size and scope but this conceptualisation. If the governments that are generous with support for the conduct of such studies as TIMSS fail to support the further analysis that is required to obtain ideas for practical improvement, then the original investments can only pay limited dividends.

5.   REFLECTIONS ON INTERNATIONAL ASSESSMENTS AND THE ISSUE OF PURPOSE

Diversity of purpose in research is so fundamental that it must be discussed and agreed in advance if a study is to bring the results desired by everyone involved. In the case of TIMSS, particularly because after the change in leadership and direction in 1993, this fundamental purpose was rarely if ever discussed generally. One view might be that it could not be discussed if all the participants – each with their own sense of the purpose of the study – were to keep on working together productively. Given the diversity that undoubtedly existed, debate about purpose at some midpoint might have risked the entire enterprise. Nonetheless, it should not be surprising if there is disappointment, even resentment, in some quarters about the final directions that TIMSS seems to have taken and therein lie some possible lessons for international assessments in the future.

It has been suggested to me privately that the reality of TIMSS is that 'he who pays the piper calls the tune' and that the comparative league tables were desired by the major financial supporters of TIMSS: the US Department of Education, National Centre for Education Statistics and the US National Science Foundation. If this is indeed the case, then IEA needs to review carefully the sources of funding for its studies and the substantive implications of these sources. It needs also to be clear and forthright with subject matter specialists, researchers and teachers whose professional time and commitment it solicits about what the goals of future studies really are. Similarly, researchers and subject-matter specialists, who give significant portions of their time to participate in these studies (usually for no personal compensation), need to be clear when they commit themselves exactly what the purpose of the enterprise is and what the results are likely to be.

Regardless of the personal aspect, international assessments require the co-operation of teachers and subject matter specialists to ensure (at a minimum) the validity of the tests. And if these people are to *want* to participate, then their legitimate interests in the outcomes of these studies must be taken seriously. It seems possible that studies in the future can be designed to make the improvement of teaching and learning the primary goal rather than a subsidiary one after the international league tables. Schmidt and McKnight are right when they point out that TIMSS was originally designed that way. The challenge now is to find the resources and people to see if this vision of TIMSS can yet be realised.

6. CONCLUDING COMMENTS

I have recounted my experience of TIMSS from a quite personal perspective and, in doing so, I have not hidden my own vision for TIMSS and, by implication, my own disappointments in aspects of the reality. However, my comments should not be interpreted as implying that I believe that the TIMSS Study Centre staff have done anything other than a quite extraordinary job of managing this most complex of studies, including its data collection, analysis and reporting. The competence, commitment, and integrity of those involved is beyond question, in my view. The tension between the subject matter specialists, whose work began the study, and the measurement specialists, whose work completed it, is their underlying vision of the purpose of TIMSS.

In conclusion, four dilemmas emerge from my experience of TIMSS that continue to puzzle me and, since the issues I have raised here were not entirely resolved within TIMSS and will reappear in future international studies, I set them out here to stimulate and provoke further debate.

– What are realistic visions for international assessments and how broadly are they shared?
– How should the trade-offs between validity and reliability be balanced to achieve the vision of those involved?
– Should 'league tables' showing rankings of participating countries be seen as useful ends in themselves, as necessary means to other useful ends, or as obstacles to the achievement of useful ends?
– If 'he who pays the piper calls the tune' how can the vision for international assessment of the education community be balanced with that of government?

As we all embark on new ventures in international assessment (such as the OECD/PISA project discussed in the following chapter), the need to reach mutually satisfactory resolution of these dilemmas becomes of ever increasing importance.

REFERENCES

Adams R .J. and Gonzalez E. J. (1996) The TIMSS Test Design, in Martin, M. O and Kelly, D. L. *The Third International Mathematics and Science Study, Technical Report, Volume 1: Design and Development,* ch. 3 (Chestnut Hill, MA: Boston College)

Beaton, A. E. (1996) Preface, in Robitaille, D. F. and Garden, R. A. *Research Questions and Study Design,* TIMSS Monograph No. 2, p. 11 (Vancouver: Pacific Educational Press)

Garden, R. A. and Orpwood, G. (1996) Development of the TIMSS Achievement Tests, in Martin, M. O and Kelly, D. L. (1996) *The Third International Mathematics and Science Study, Technical Report, Volume 1: Design and Development,* ch. 3 (Chestnut Hill, MA, Boston College)

Mullis, I. V. S., Martin, M. O., Beaton, A. E., Gonzalez, E. J., Kelly, D. L., and Smith, T. A. (1998) *Mathematics and Science Achievement in the Final Year of Secondary School: IEA's Third International Mathematics and Science Study (TIMSS)* (Chestnut Hill, MA: Boston College)

Orpwood, G. (ed.) (1998) *Ontario in TIMSS: Secondary Analysis and Recommendations,* Unpublished report prepared on behalf of the Ontario Association of Deans of Education for the (Ontario) Educational Quality and Accountability Office

Orpwood, G. and Garden, R. A. (1998) *Assessing Mathematics and Science Literacy.* TIMSS Monograph No. 4 (Vancouver: Pacific Educational Press)

Robitaille, D. F. and Robeck, E. C. (1996) The Character and the Context of TIMSS, in Robitaille, D. F. and Garden, R. A. (eds.). *Research Questions and Study Design,* TIMSS Monograph No. 2, p. 15 (Vancouver: Pacific Educational Press)

Schmidt, W. H. and McKnight, C. C. (1998) So What Can We Really Learn from TIMSS? Paper published on the Internet (http://TIMSS.msu.edu/whatcanwelearn.htm)

Schmidt, W. H., McKnight, C. C., Valverde, G. A., Houang, R. T., and Wiley, D. E. (1997) *Many Visions, Many Aims, Volume 1: A Cross-National Investigation of Curricular Intentions in Mathematics* (Dordrecht: Kluwer)

Schmidt, W. H, Raizen, S. A., Britton, E. D., Bianchi, L. J, and Wolfe, R. G. (1997) *Many Visions, Many Aims, Volume 2: A Cross-National Investigation of Curricular Intentions in Science* (Dordrecht: Kluwer)

Chapter 4

Monitoring Student Knowledge and Skills
The OECD Programme for International Student Assessment

Andreas Schleicher
OECD/SEELSA Statistics and Indicators Division, Paris

1.   INTRODUCTION

How well are young adults prepared to meet the challenges of the future? Are they able to analyse, reason and communicate their ideas effectively? Do they have the capacity to continue learning throughout life? Parents, students, the public and those who run education systems need to know.

Many education systems monitor student learning to provide some answers to these questions. Comparative international analyses can extend and enrich the national picture by establishing the levels of performance being achieved by students in other countries and by providing a larger context within which to interpret national results. They can provide direction for schools' instructional efforts and for students' learning as well as insights into curriculum strengths and weaknesses. Coupled with appropriate incentives, they can motivate students to learn better, teachers to teach better and schools to be more effective. They also provide tools to allow central authorities to monitor achievement levels even when administration is devolved and schools are being run in partnership with communities.

Governments and the general public need solid and internationally comparable evidence of educational outcomes. In response to this demand, the Organisation for Economic Co-operation and Development (OECD) has launched the Programme for International Student Assessment (PISA). OECD/PISA will produce policy-oriented and internationally comparable indicators of student achievement on a regular and timely basis. The assessments will focus on 15 year-olds and the indicators are designed to

D. Shorrocks-Taylor and E.W. Jenkins (eds.), Learning from Others, 63–77.

contribute to an understanding of the extent to which education systems in participating countries are preparing their students to become lifelong learners and to play constructive roles as citizens in society.

OECD/PISA represents a new commitment by the governments of OECD countries to monitor the outcomes of education systems in terms of student achievement, within a common framework that is internationally agreed. While it is expected that many individuals in participating countries, including professionals and lay persons, will use the survey results for a variety of purposes, the primary reason for developing and conducting this large-scale international assessment is to provide empirically grounded information which will inform policy decisions.

The results of the OECD assessments, to be published every three years along with other indicators of education systems, will allow national policy makers to compare the performance of their education systems with those of other countries. They will also help to focus and motivate educational reform and school improvement, especially where schools or education systems with similar inputs achieve markedly different results. Further, they will provide a basis for better assessment and monitoring of the effectiveness of education systems at the national level.

OECD/PISA is a collaborative process. It brings together scientific expertise from the participating countries and is steered jointly by their governments on the basis of shared, policy-driven interests.

2.   WHAT IS OECD/PISA?
A SUMMARY OF KEY FEATURES

2.1   Basics

– OECD/PISA is an internationally standardised assessment, jointly developed by participating countries and administered to 15 year-olds in groups in their schools.
– The study will be administered in 32 countries, of which 28 are members of the OECD, and between 4,500 and 10,000 students will typically be tested in each country.

2.2   Content

– PISA covers three domains: reading literacy, mathematical literacy and scientific literacy.
– PISA aims to define each domain not merely in terms of mastery of the school curriculum, but in terms of important knowledge and skills needed

in adult life. The assessment of cross-curriculum competencies is an integral part of PISA.
– Emphasis is placed on the mastery of processes, the understanding of concepts and the ability to function in various situations within each domain.

2.3   Methods

– Pencil and paper tests are used, with assessments lasting a total of two hours for each student.
– Test items are a mixture of multiple-choice test items and questions requiring the student to construct their own responses. The items are organised in groups based on a passage setting out a real-life situation.
– A total of about seven hours of test items is included, with different students taking different combinations of the test items.
– Students answer a background questionnaire which takes 20-30 minutes to complete, providing information about themselves. School principals are given a 30-minute questionnaire asking about their schools.

2.4   Assessment cycle

– The first assessment will take place in 2000, with first results published in 2001, and will continue thereafter in three-year cycles.
– Each cycle looks in depth at a 'major' domain, to which two-thirds of testing time are devoted; the other two domains provide a summary profile of skills. Major domains are reading literacy in 2000, mathematical literacy in 2003 and scientific literacy in 2006.

2.5   Outcomes

– A basic profile of knowledge and skills among students at the end of compulsory schooling.
– Contextual indicators relating results to student and school characteristics.
– Trend indicators showing how results change over time.

3.   BASIC FEATURES OF OECD/PISA

Although the domains of· reading literacy, mathematical literacy and scientific literacy correspond to school subjects, the OECD assessments will not primarily examine how well students have mastered the specific curriculum content. Rather, they aim at assessing the extent to which young people have acquired the wider knowledge and skills in these domains that they will need in adult life. The assessment of cross-curricular competencies has, therefore, been made an integral part of OECD/PISA. The most important reasons for this broadly oriented approach to assessment are as follows.

First, although specific knowledge acquisition is important in school learning, the application of that knowledge in adult life depends crucially on the individual's acquisition of broader concepts and skills. In reading, the capacity to develop interpretations of written material and to reflect on the content and qualities of text are central skills. In mathematics, being able to reason quantitatively and to represent relationships or dependencies is more important than the ability to answer familiar textbook questions when it comes to deploying mathematical skills in everyday life. In science, having specific knowledge, such as the names of specific plants and animals, is of less value than an understanding of broad concepts and topics such as energy consumption, biodiversity and human health in thinking about the issues of science under debate in the adult community.

Second, a focus on curriculum content would, in an international setting, restrict attention to curriculum elements common to all, or most, countries. This would force many compromises and result in an assessment that was too narrow to be of value for governments wishing to learn about the strengths and innovations in the education systems of other countries.

Third, there are broad, general skills that it is essential for students to develop. These include communication, adaptability, flexibility, problem solving and the use of information technologies. These skills are developed across the curriculum and an assessment of them requires a cross-curricular focus.

Underlying OECD/PISA is a dynamic model of lifelong learning in which new knowledge and skills necessary for successful adaptation to changing circumstances are continuously acquired over the life cycle. Students cannot learn in school everything they will need to know in adult life. What they must acquire is the prerequisites for successful learning in future life. These prerequisites are of both a cognitive and a motivational nature. Students must become able to organise and regulate their own learning, to learn independently and in groups, and to overcome difficulties in the learning process. This requires them to be aware of their own thinking processes and learning strategies and methods. Moreover, further learning and the acquisition of additional knowledge will increasingly occur in situations in which people work together and are dependent on one another. To assess these aspects, the development of an instrument that seeks information on self-regulated learning is being explored as part of the OECD/PISA 2000 assessment.

OECD/PISA is not a single cross-national assessment of the reading, mathematics and science skills of 15 year-olds. It is an on-going programme of assessment that will gather data from each of these domains every three years. Over the longer term, this will lead to the development of a body of information for monitoring trends in the knowledge and skills of students in the various countries as well as in different demographic sub-groups of each country. On each occasion, one domain will be tested in detail, taking up nearly two-thirds of the total testing time. This 'major' domain will be reading literacy in 2000, mathematical literacy in 2003 and scientific literacy in 2006. This cycle will provide a thorough analysis of achievement in each area every nine years, and a 'check-up' every three.

The total time spent on the tests by each student will be two hours but information will be obtained on almost seven hours' worth of test items. The total set of questions will be packaged into a number of different groups. Each group will be taken by a sufficient number of students for appropriate estimates to be made of the achievement levels on all items by students in each country and in relevant sub-groups within a country (such as males and females and students from different social and economic contexts). Students will also spend 20 minutes answering questions for the context questionnaire.

The assessments will provide various types of indicators:
– basic indicators providing a baseline profile of the knowledge and skills of students;
– contextual indicators, showing how such skills relate to important demographic, social, economic and educational variables;
– indicators on trends that will emerge from the on-going, cyclical nature of the data collection and that will show changes in outcome levels, changes in outcome distributions and changes in relationships between student-level and school-level background variables and outcomes over time.

Although indicators are an adequate means of drawing attention to important issues, they are not usually capable of providing answers to policy questions. OECD/PISA has therefore also developed a policy-oriented analysis plan that will go beyond the reporting of indicators.

The countries participating in the first OECD/PISA survey cycle are: Australia, Austria, Belgium, Brazil, Canada, China, the Czech Republic, Denmark, Finland, France, Germany, Greece, Hungary, Iceland, Ireland, Italy, Japan, Korea, Latvia, Luxembourg, Mexico, the Netherlands, New Zealand, Norway, Poland, Portugal, the Russian Federation, Spain, Sweden, Switzerland, the United Kingdom and the United States.

4.   HOW OECD/PISA IS DIFFERENT FROM OTHER INTERNATIONAL ASSESSMENTS

OECD/PISA is not the first international comparative survey of student achievement. As this volume makes clear, others have been conducted over the past 40 years, primarily by the International Association for the Evaluation of Educational Achievement (IEA) and by the Education Testing Service's International Assessment of Educational Progress (IAEP). The quality and scope of these surveys have greatly improved over the years but they provide only partial and sporadic information about student achievement in limited subject areas. The three science and mathematics surveys conducted by the IEA provide some indication of how things have changed over 30 years but the view is limited by the restricted numbers of countries participating in the early surveys and by limitations on the extent to which the tests can be compared.

More importantly, though, these surveys have concentrated on outcomes linked directly to the curriculum and then only on those parts of the curriculum that are essentially common across the participating countries. Aspects of the curriculum unique to one country or a small number of countries have usually not been taken into account in the assessments, regardless of how significant that part of the curriculum is for the countries involved.

OECD/PISA is taking a different approach in a number of important respects which make it distinctive. These are:
–   its origin: it is governments that have taken the initiative and whose policy interests the survey will be designed to serve,
–   its regularity: the commitment to cover multiple assessment domains with updates every three years will make it possible for countries to regularly and predictably monitor their progress in meeting key learning objectives, and
–   the age-group covered: assessing young people near the end of their compulsory schooling gives a useful indication of the performance of education systems. While most young people in OECD countries continue their initial education beyond the age of 15, this is normally close to the end of the initial period of basic schooling in which all young people follow a broadly common curriculum. It is useful to determine, at that stage, the extent to which they have acquired knowledge and skills that will help them in the future, including the individualised paths of further learning which they may follow.
–   The knowledge and skills tested: these are defined not primarily in terms of a common denominator of national school curricula but in terms of what skills are deemed to be essential for future life. This is the most

fundamental and ambitious novel feature of OECD/PISA. It would be arbitrary to draw too precise a distinction between 'school' skills and 'life' skills, since schools have always aimed to prepare young people for life, but the distinction is important. School curricula are traditionally constructed largely in terms of bodies of information and techniques to be mastered. They traditionally focus less, within curriculum areas, on the skills to be developed in each domain for use generally in adult life. They focus even less on more general competencies, developed across the curriculum, to solve problems and apply one's ideas and understanding to situations encountered in life. OECD/PISA does not exclude curriculum-based knowledge and understanding, but it tests for it mainly in terms of the acquisition of broad concepts and skills that allow knowledge to be applied. Further, OECD/PISA is not constrained by the common denominator of what has been specifically taught in the schools of participating countries.

This emphasis on testing in terms of mastery and broad concepts is particularly significant in the light of the concern among nations to develop human capital, which the OECD defines as:

"the knowledge, skills, competencies and other attributes embodied in individuals that are relevant to personal, social and economic well-being".

Estimates of the stock of human capital or human skill base have tended, at best, to be derived using proxies such as level of education completed. When the interest in human capital is extended to include attributes that permit full social and democratic participation in adult life and that equip people to become 'lifelong learners', the inadequacy of these proxies becomes even clearer.

By directly testing for knowledge and skills close to the end of basic schooling, OECD/PISA examines the degree of preparedness of young people for adult life and, to some extent, the effectiveness of education systems. Its ambition is to assess achievement in relation to the underlying objectives (as defined by society) of education systems, not in relation to the teaching and learning of a body of knowledge. Such authentic outcome measures are needed if schools and education systems are to be encouraged to focus on modern challenges.

5.   WHAT IS BEING ASSESSED IN EACH DOMAIN?

Table 4.1 summarises the structure of each of the three OECD/PISA domains, the definition of each domain and the dimensions that characterise the test items.

Table 4.1. Summary of PISA dimensions

	Domain		
	Reading literacy	Mathematical literacy	Scientific literacy
Definition	Understanding, using and reflecting on written texts, in order to achieve one's goals, to develop one's knowledge and potential, and to participate in society.	Identifying, understanding and engaging in mathematics and making well-founded judgements about the role that mathematics plays, as needed for an individual's current and future life as a constructive, concerned and reflective citizen.	Combining scientific knowledge with the drawing of evidence-based conclusions and developing hypotheses in order to understand and help make decisions about the natural world and the changes made to it through human activity.
Components/ dimensions of the domain	Reading different kinds of text: continuous prose sub-classified by type (e.g. description, narration) and documents, sub-classified by structure.

Performing different kinds of reading tasks, such as retrieving specific information or demonstrating a broad under-standing of text.

Reading texts written for different situations, e.g. for personal interest, or to meet work requirements | Mathematical content – primarily mathematical 'big ideas', in the first cycle these are i) change and growth and ii) space and shape. In future cycles also chance, quantitative reasoning, uncertainty and dependency relationships. Also curricular strands such as number, algebra will be covered, but are less significant than 'big ideas'.

Mathematical competencies, e.g. modelling, problem-solving; divided into three classes – i) carrying out procedures, ii) making connections and iii) mathematical thinking/ insight

Using mathematics in different situations, e.g. problems that affect individuals, communities or the whole world. | Scientific concepts – e.g. energy conservation, adaptation, decomposition – chosen from the major fields of physics, biology, chemistry etc, as being relevant to everyday situations, where they are applied in matters to do with the use of energy, the maintenance of species or the use of materials.

Process skills – e.g. identifying evidence, communicating valid conclusions. These do not depend on a pre-set body of scientific knowledge, but cannot be applied in the absence of scientific content.

Situations or context of science use, e.g. problems that affect *individuals*, communities or the whole world. |

The definitions of all three domains place an emphasis on functional knowledge and skills that allow one to participate actively in society. Such participation requires more than just being able to carry out tasks imposed externally by, for example, an employer. It also means being equipped to take part in decision-making processes. In the more complex tasks in OECD/PISA, students will be asked to reflect on material, not just to answer questions that have single 'correct' answers.

In order to operationalise these definitions, each domain is described in terms of three dimensions. These correspond roughly to:
- the content or structure of knowledge that students need to acquire in each domain;
- a range of processes that need to be performed, which require various cognitive skills; and
- the situation or context in which knowledge and skills are applied or drawn on.

The idea is to assess students across a range of skills required for a variety of tasks that citizens are likely to encounter. However, it should be emphasised that the relative importance of the three dimensions still needs to be explored. Field trials in 1999 assessed a large number of questions on the three dimensions, with varying characteristics, before a decision was reached about the most useful range of characteristics and the manner in which to translate performance in various items into aggregate scores.

Within the common framework of the three dimensions, each domain defines its various dimensions in particular ways. There is an important difference between reading literacy on the one hand, and science and mathematics on the other. The former is itself a skill that cuts across the curriculum, particularly in secondary education, and does not have any obvious 'content' of its own. Although some formal understanding of structural features such as sentence structure may be important, such knowledge cannot be compared, for example, to the mastery of a range of scientific principles or concepts.

The following are the main aspects of each of the three assessment domains summarised in Table 4.1.

Reading literacy is defined in terms of individuals' ability to use written text to achieve their purposes. This aspect of literacy has been well established by previous surveys such as the International Adult Literacy Survey (IALS), but is taken further in OECD/PISA by the introduction of an 'active' element – the capacity not just to understand a text but to reflect on it, drawing on one's own thoughts and experiences. Reading literacy is assessed in relation to:
1. The form of reading material, or text. Many student reading assessments have focused on 'continuous texts' or prose organised in sentences and

paragraphs. OECD/PISA will in addition introduce 'non-continuous texts' which present information in other ways, such as in lists, forms, graphs, or diagrams. It will also distinguish between a range of prose forms, such as narration, exposition and argumentation. These distinctions are based on the principle that individuals will encounter a range of written material in adult life, and that it is not sufficient to be able to read a limited number of text types typically encountered in school.

2. The type of reading task. This corresponds at one level to the various cognitive skills that are needed to be an effective reader, and at another, to the characteristics of questions set in the assessment. Students will not be assessed on the most basic reading skills, as it is assumed that most 15 year-olds will have acquired these. Rather, they will be expected to demonstrate their proficiency in retrieving information, forming a broad general understanding of the text, interpreting it, reflecting on its contents and its form and features.

3. The use for which the text was constructed – its context or situation. For example, a novel, personal letter or biography is written for people's 'personal' use; official documents or announcements for 'public' use; a manual or report for 'occupational' use; and a textbook or worksheet for 'educational' use. An important reason for making these distinctions is that some groups may perform better in one reading situation than in another, in which case it is desirable to include a range of types of reading in the assessment items.

Mathematical literacy is defined in terms of the individual's understanding of the role of mathematics and the capacity to engage in this discipline in ways that meet his or her needs. This puts the emphasis on the capacity to pose and solve mathematical problems rather than to perform specified mathematical operations. Mathematical literacy is assessed in relation to:

1. The content of mathematics, as defined mainly in terms of mathematical 'big ideas' (such as chance, change and growth, space and shape, quantitative reasoning, uncertainty and dependency relationships) and only secondarily in relation to 'curricular strands' (such as numbers, algebra and geometry). A representative, rather than a comprehensive, range of the main concepts underlying mathematical thinking have been chosen for OECD/PISA; these have been narrowed down further for the first cycle of the assessment – in which mathematics is a 'minor' domain – to two big ideas: change and growth and space and shape. These allow a wide representation of aspects of the curriculum without undue focus on number skills.

2. The process of mathematics as defined by general mathematical competencies. These include the use of mathematical language, modelling and problem-solving skills. The idea is not, however, to separate out such skills in different test items, since it is assumed that a range of competencies will be needed to perform any given mathematical task. Rather, questions are organised in terms of three 'competency classes' defining the type of thinking skill needed. The first class consists of simple computations or definitions of the type most familiar in conventional mathematics assessments. The second requires connections to be made to solve straightforward problems. The third competency class consists of mathematical thinking, generalisation and insight, and requires students to engage in analysis, to identify the mathematical elements in a situation and to pose their own problems.

3. The situations in which mathematics is used. The framework identifies five situations: personal, educational, occupational, public and scientific. In the case of mathematics, this dimension is considered to be less important than process or content.

Scientific literacy is defined in terms of being able to use scientific knowledge and processes not just to understand the natural world but to participate in decisions that affect it. Scientific literacy is assessed in relation to:

1. Scientific concepts, which constitute the links aiding understanding of related phenomena. In OECD/PISA, whilst the concepts are the familiar ones relating to physics, chemistry, biological sciences and earth and space sciences, they will need to be applied to the content of the items and not just recalled. The main item content will be selected from within three broad areas of application: science in life and health, science in earth and environment and science in technology.

2. Scientific processes, centred on the ability to acquire, interpret and act upon evidence. Five such processes present in OECD/PISA relate to: i) the recognition of scientific questions, ii) the identification of evidence, iii) the drawing of conclusions, iv) the communication of these conclusions and v) the demonstration of understanding of scientific concepts. Only the last of these requires a pre-set body of scientific knowledge, yet since no scientific process can be 'content free', the OECD/PISA science questions will always require an understanding of concepts in the areas mentioned above.

3. Scientific situations, selected mainly from people's everyday lives rather than from the practice of science in a school classroom or laboratory, or the work of professional scientists. The science situations in OECD/ PISA are defined as matters relating to the self and the family, to the

wider community, to the whole world and to the historical evolution of scientific understanding.

A detailed description of the assessment domains and their operationalisation for the purpose of the OECD/PISA tests can be found in *Measuring Student Knowledge and Skills – A new Framework for Assessment*, OECD, 1999.

6.   HOW THE ASSESSMENT WILL TAKE PLACE AND HOW RESULTS WILL BE REPORTED

For reasons of feasibility, OECD/PISA 2000 will consist of pencil and paper instruments. Other forms of assessments will be actively explored for subsequent cycles.

The assessment will be made up of items of a variety of types. Some items will be 'closed' – that is, they will require students to select or produce simple responses that can be directly compared with a single correct answer. Others will be more open, requiring students to provide more elaborate responses designed to measure broader constructs than those captured by other, more traditional surveys. The assessment of higher-order skills, often through open-ended problems, will be an important innovation in OECD/PISA. The extent to which this type of exercise will be used will depend on how robust the methodology proves to be and how consistent a form of marking can be developed. The use of open-ended items is likely to grow in importance in successive OECD/PISA cycles, from a relatively modest start in the first survey cycle.

In most cases the assessments will consist of sets of items relating to a common text, stimulus or theme. This is an important feature that allows questions to go into greater depth than would be the case if each question introduced a wholly new context. It allows time for the student to digest material that can then be used to assess multiple aspects of performance.

Overall, the items in OECD/PISA will look quite different from those used, for example, in earlier international assessments such as the Third International Mathematics and Science Study (IEA/TIMSS), which concentrated on short multiple-choice questions based on what had been learned at school. For example, some science questions may only require straightforward knowledge (e.g. asking students to specify how many legs and body parts insects have), or the simple manipulation of knowledge (e.g. students have to work out whether a metal, wooden or plastic spoon would feel hottest after being placed in hot water for 15 seconds). OECD/PISA items, on the other hand, generally require the combination of a variety of knowledge and competencies, and sometimes an active evaluation of decisions for which there is no single correct answer.

OECD/PISA results will be reported in terms of the level of performance on scales of achievement in each domain. The calibration of the tasks in the tests on to scales will provide a language with which to describe the competencies exhibited by students performing at different levels. That is, it will be possible to say what students at any level on each scale know and are able to do that those below them cannot. The inclusion of items that require higher order thinking skills, as well as others that entail relatively simple levels of understanding, will ensure that the scales cover a wide range of competencies.

An important issue will be whether the levels of literacy in each domain should be reported on more than one scale. Can a person's competencies be easily aggregated and placed on a specific level, or is it more useful to describe them as having reached different levels in relation to different aspects? That will depend on two things: first, the extent to which an individual's performance in one kind of question correlates with performance in another, and the patterns of any differences in performance across particular dimensions; and second, the feasibility, given the number of items that can be included in the assessment, of reporting on more than one scale in each domain. Each scale corresponds to one type of score that will be assigned to students; having more than one scale therefore implies that students will be assigned multiple scores reflecting different aspects of the domain. The most likely scenario is that in the major domain that contains the most questions (reading in OECD/PISA 2000) there will be scope for more than one scale but in the minor domains a single scale will be used.

7.   THE CONTEXT QUESTIONNAIRES AND THEIR USE

To gather contextual information, students and the principals of their schools will also be asked to respond to background questionnaires that will take 20 to 30 minutes to complete. These questionnaires are central tools to allow the analysis of results in terms of a range of student and school characteristics.

The questionnaires will seek information about:
- the students and their family backgrounds, including the economic, social and cultural capital of students and their families;
- aspects of students' lives such as their attitudes to learning, their habits and life inside school and their family environment;

– aspects of schools such as the quality of the school's human and material resources, public and private control and funding, decision-making processes and staffing practices;
– the context of instruction including institutional structures and types, class size and the level of parental involvement.

The first cycle of OECD/PISA will also include an instrument that asks students to report on self-regulated learning. This instrument is based on the following components:

– strategies of self-regulated learning, which govern how deeply and how systematically information is processed;
– motivational preferences and goal orientations, which influence the investment of time and mental energy for learning purposes as well as the choice of learning strategies;
– self-related cognition mechanisms, which regulate the aims and processes of action;
– action control strategies, particularly effort and persistence, that protect the learning process from competing intentions and help to overcome learning difficulties;
– preferences for different types of learning situations, learning styles and social skills required for co-operative learning.

Overall, the OECD/PISA context questionnaires will provide a detailed basis for policy-oriented analysis of the assessment results. Together with information obtained through other OECD instruments and programmes, it will be possible to:

– compare differences in student outcomes with variations in education systems and the instructional context;
– compare differences in student outcomes with differences in curriculum content and pedagogical processes;
– consider relationships between student performance and school factors such as size and resources, as well as the differences between countries in these relationships;
– examine differences between countries in the extent to which schools moderate or increase the effects of individual-level student factors on student achievement;
– consider differences in education systems and the national context that are related to differences in student achievement between countries.

The contextual information collected through the student and school questionnaires will comprise only a part of the total amount of information available to OECD/PISA. Indicators describing the general structure of the education systems (their demographic and economic contexts, for example, costs, enrolments, throughput, school and teacher characteristics, and some classroom processes) and on labour market outcomes, are already routinely developed and applied by the OECD.

8.   OECD/PISA – AN EVOLVING INSTRUMENT

Given the long horizon of the project and the different relative emphases that will be given to the domains in each cycle, the OECD/PISA assessment frameworks clearly represent an instrument that will evolve.   The frameworks are designed to be flexible so that they can evolve and adapt to the changing interests of participants; and yet give guidance to those enduring elements that are likely to be of continuing concern and therefore should be included in all assessment cycles.

The frameworks have been modified in the light of field trials in the course of 1999 and a final instrument produced for use in OECD/PISA 2000. Moreover, the development of the survey will continue thereafter, to take account of both changing objectives of education systems and improvements in assessment techniques.   The benefits of such development and improvement will, of course, have to be balanced against the need for reliable comparisons over time, so that many of the core elements of OECD/PISA will be maintained over the years.

OECD's objectives are ambitious.   For the first time an international assessment of school students aims to determine not just whether they have acquired the knowledge specified in the school curriculum, but whether the knowledge and skills that they have acquired in childhood have prepared them well for adult life.   Such outcome measures are needed by countries which want to monitor the adequacy of their education systems in a global context.   The ideal will not be instantly achieved, and some of OECD/PISA's goals will initially be constrained by what is practicable in an assessment instrument that needs also to be reliable and comparable across many different cultures.   But the objectives are clear, and the assessments that evolve over the coming years will aim to move progressively towards fulfilling them.

Chapter 5

International Comparisons – What really matters?[5.1]

Julian Elliott and Neil Hufton
University of Sunderland, UK

Leonid Illushin
Hertzen University, St Petersburg, Russia

1. INTRODUCTION

International comparisons of educational achievement such as the Third International Mathematics and Science Study (Beaton *et al.*, 1996a; 1996b and Chapter 2 in this volume) have resulted in recommendations from highly influential quarters in the UK that we should look to the East to gain insights as to how best to raise educational performance. The focus of our attentions, it has been widely argued, should be pedagogic practices that essentially consist of various forms of whole class teaching. Classroom practices in Eastern Europe (Burghes, 1996) and Taiwan (Reynolds and Farrell, 1996) have been seen to be key in maintaining high standards of achievement and such perspectives have led to a reconsideration of classroom organisation and pedagogical approaches (e.g. Prais, 1997; Politeia 2000; Whitburn 2000) which has, in turn, informed the operation of national programmes for literacy and numeracy in primary schools.

The key issue is, however, whether observers of high achieving countries are correctly identifying the reasons for such performance and can determine what actually matters when ideas and practices are imported into new contexts. What tends to be overlooked is that whole class teaching is not only the preserve

---

[5.1] This paper is substantially the same as that published in British Education Research Journal, Vol 25, No 1, pp. 75-94, although additional interview data has now been added.

D. Shorrocks-Taylor and E.W. Jenkins (eds.), Learning from Others, 79–114.

of those countries that score highly in international comparative studies but is, in actuality, practised almost universally. Surely, there are other rather more intangible processes which underpin high performance? In this chapter, it will be argued that the current world-wide preoccupation with methods of curriculum delivery and teaching style may mask more fundamental reasons for educational underachievement. Crucial to this is the motivation to work hard which, while having some roots in school and classroom dynamics (Hufton and Elliott, 2000), is greatly influenced by wider sociocultural contexts.

Despite its strong academic tradition, widespread concern about educational attainment is also currently being voiced in Russia. Massive social and economic upheaval appear to be leading to a lessening of the esteem in which education has been traditionally held, student motivation is declining and drop-out rates have soared (Prelovskaya, 1994). Although international studies suggest that the performance of Russian children, and of those in other developed Eastern European countries, continues to exceed that of the UK and USA (Foxman, 1992; Beaton *et al.*, 1996a, 1996b), concern is being expressed by Russian educationalists that their standards may not be maintained in the future. Economic instability has not only resulted in a growth of materialism and a concomitant devaluation of academic scholarship but also a reduction in educational expenditure which is currently three to four times less than in other European countries and six times less than in the USA (Chuprov and Zubok, 1997). In response, new teaching methods are being introduced from Western Europe and the United States in a relatively haphazard, school by school fashion, in the hope that these will arrest any decline in performance (Bridgman, 1994). While some teachers have welcomed this new-found freedom to innovate, others have become confused and unsure about how to respond to demands that they become the agents of change (Webber, 1996). To the observer, however, it is unclear to what extent Western teaching and learning techniques are suitable for East European cultures (Burton, 1997), and, even if they are, whether their importation can offset the demotivating impact of socio-economic transformation.

Concern about the poor academic performance of many American children is a less recent phenomenon (Howley *et al.*, 1995). More recently, however, this has been fuelled by findings (US Department of Education, 1998) that US children's comparative international position in science and mathematics deteriorates as the cohort under examination shifts from fourth to eighth to twelfth grade. This trend appears to reflect the views of many US commentators that an attitudinal/motivational malaise adversely affects the performance of high school students.

In the State of Kentucky, one of the educationally less-successful areas of the United States, attempts have been made to raise performance by the introduction of the Kentucky Education Reform Act. A central plank of this legislation is a focus upon changing classroom practice, which requires teachers to adopt many of those very child-centred practices which have been recently condemned in Britain. A shift towards activity-based problem solving rooted in experience, with a concomitant reduction in traditional classroom teaching, is virtually mandatory of Kentucky primary school teachers.

The current world-wide preoccupation with methods of curriculum delivery and teaching style may mask more fundamental reasons for educational underachievement. In comparing Asian with Western achievement and classroom practice, it is often overlooked that there is a large American literature which indicates that US-based Asian American children consistently outperform their Caucasian American counterparts (Hirschman and Wong, 1986; Eaton and Dembo, 1997) even when controlling for ability (Steinberg, Dornbusch and Brown, 1992) and socio-economic status (Ogbu, 1983). As this can't be explained by differences in classroom practices, this finding supports the oft-quoted claim that the higher performance of Asian children is rooted in attitudinal and motivational differences originating in familial and (sub)cultural contexts.

Supporting evidence for the central importance of sociocultural factors is provided by Robinson (1997) who in a longitudinal study of children born in 1970, found that such variables as pre-school education, class size, teaching method, homework policy and streaming had a very limited role in determining literacy and numeracy skills. Far more important were social class, parental interest and peer group pressure — factors which are likely to have a significant impact upon children's attitudes towards, and expectations of, school.

It has been widely argued that many children, particularly those from economically disadvantaged communities, are insufficiently challenged in school (OFSTED, 1993). In part, this practice may result from recognition of the need to ensure children's sense of personal competence and self-worth. Oettingen, Little, Lindenberger and Baltes (1994) have pointed to a Western tendency to encourage high self-esteem in the face of poor academic performance. This may lead to an apparent acceptance of, and satisfaction with, mediocre performance. Compounding this phenomenon is the strong desire for peer approval, which, for many children, may reduce academic endeavour: in many Western school cultures, intellectual enthusiasm is clearly inimical to social acceptance (Howley *et al.*, 1995; Steinberg, 1996). Public recognition that an individual lacks intellectual ability is often a matter of shame, however, and this can lead the child to disengage from academic activity as making a determined effort in class and then failing may attract the attribution of low ability (Covington, 1992). Clearly, children's understandings of the nature of intelligence and their attributions for success and failure are likely to have a significant impact upon motivation and academic achievement (Dweck, 1986: Stipek and Gralinski, 1996).

In a series of studies over several years, a team of American and Asian researchers (Stevenson and Stigler, 1992; Stevenson, and Lee, 1990) reported significant differences in the attitudes to education of children, teachers and parents in the United States, Taiwan, Japan and China. In comparing the educational performance of a sample of Chinese, Japanese and American children in elementary schooling, Stevenson and Stigler (1992) noted dramatic discrepancies in mathematical performance at fifth grade to the extent that the highest performing American school was poorer than the lowest performing Asian schools. Only one American child was placed within the top one hundred scorers.

In pinpointing reasons for the higher Asian performance, these researchers outlined a number of important factors which appeared to be independent of school or classroom organisation. Many of these centred upon family pressures to strive educationally. Asian parents were far less satisfied with mediocre performance than their American counterparts and their active encouragement helped to ensure that their children spent a significantly higher proportion of time studying during out of school hours.

In considering the topic of motivation, perhaps the most influential perspective during the past decade has been the effect of attributional biases upon goal seeking behaviour. The seminal work of Weiner (1979) suggested that where attributions for success were external to the individual (e.g. task difficulty or luck), motivation to persist was reduced. In contrast, internal attributions, where the learner attributes success or failure to their own characteristics, are more likely to result in increased motivation to succeed. In their comparative study, Stevenson and Stigler (1992) stressed the importance of student attributions for success and failure as key to motivation and educational performance. They report the findings of several studies demonstrating the stronger emphasis that Asian children placed upon effort as the key to success. In contrast, US children appeared to emphasise fixed ability as an important factor. Steinberg (1996) reports similar findings in a series of studies of US high school students where children from Asian families were found to have a more desirable attributional style (i.e. one emphasising effort over ability) than their Black, Latino or White peers.

Recent studies suggest that English children tend to attribute success in school to internal rather than external factors (Blatchford, 1996; Lightbody *et al.,* 1996; Gipps and Tunstall, 1998). An internal/external differentiation by itself is of limited value, for this can be further divided into those aspects over which the individual has major control, for example, effort, and those such as ability which may be assumed to be largely fixed and uncontrollable (Weiner *op. cit.*). Children who attribute academic failure to ability rather than effort, it is argued, are less likely to persist when confronted by difficulty. The relative importance of effort and ability has been highlighted in recent considerations of British children's poor performance relative to countries in south-east Asia (Gipps, 1996; Reynolds and Farrell, 1996). In contrast to the Asian emphasis upon effort, underpinned by Confucian beliefs, Western culture, it is frequently argued, is more influenced by notions of fixed intelligence and relatively stable levels of ability.

While Blatchford's study does not appear to explore the controllability dimension explicitly, his respondents rarely appeared to offer ability as a factor influencing their own performance. Similarly, Lightbody *et al.*'s study of children in one London secondary school indicated a greater occurrence of effort, rather than ability, attributions. It is important to note, however, that in both of these studies, children could be linked to their responses. Blatchford's study involved face to face interviews while Lightbody *et al.*, required respondents to identify themselves on the questionnaire response sheet. This must raise questions about social desirability and impression management influences. As Blatchford (*op. cit.*) points out, teachers are continually stressing the importance of effort and persistence and it would be unsurprising if the children in such studies reflected such messages in their responses.

Gipps and Tunstall (1998) provided short 'stories' about classroom performance to 49 six and seven year-olds. Effort was the most commonly cited reason provided by the children when asked to give reasons for success or failure in these vignettes. Of secondary importance was competence in the specific domain under consideration. These findings, suggest Gipps and Tunstall,

> ... seem to question the oft-quoted Anglo-Saxon belief in ability (p.161).

Many American researchers, concerned by poor standards in US schools, have focused upon social pressures, which undermine academic progress. In reviewing his extensive longitudinal study of more than 20,000 high school students, together with their parents and teachers, Steinberg (1996) provides a damning indictment of contemporary American attitudes to education. His findings demonstrate large-scale parental disengagement from schooling whereby acceptance of poor grades is widespread, a peer culture that is often scornful of academic excellence, and student lifestyles in which a high proportion of time outside of school is spent socialising, engaged in leisure pursuits and/or part-time employment. In the light of his, and others' findings, Steinberg concludes that too much emphasis has been placed upon school reform, and insufficient attention given to attitudes, beliefs and achievement-striving behaviours:

> No curricular overhaul, no instructional innovation, no change in school organisation, no toughening of standards, no rethinking of teacher

training or compensation will succeed if students do not come to school interested in, and committed to, learning. (p. 194).

A number of American commentators (Bishop, 1989; Rosenbaum, 1989), have explained poor educational attainment in largely economic terms. They point out that effort and achievement in high-school are not clearly rewarded by success in the labour market as school grades and test scores are often unrelated to the gaining of prestigious jobs and salaries. Bishop (*op.cit.*) also argues that admission to selective universities is insufficiently related to school performance as selection is largely based upon aptitude tests rather than school subject performance. The culture of mediocrity, he adds, is further maintained by the use of class ranking and grade point averages in selection – judgements which are made relative to classmates rather than to any external standard. In common with almost all US researchers examining educational underachievement, Bishop (*op.cit.*) also highlights the existence of strong peer pressure against academic commitment which contrasts strongly with sporting endeavours.

Such pressures also seem to be evident in England where the presence of a peer-induced, anti-work ethic appears not to be solely an adolescent phenomenon. In a recent comparative study of the attitudes of French and English primary school children, for example, it was noted that:

> '... many English children did not want to be best in the class, and felt lukewarm about getting a good mark or even praise for good work. Some children actually said they did not want to be seen as too good by the teacher; no French child said this' (Osborn, 1997, p. 49).

In developing from, and building upon, the research findings outlined above, the present study represents the first stage of a detailed examination into the relationship between a number of key sociocultural and pedagogic factors and educational achievement in three very different regions: the city of Sunderland, in north-east England, St.Petersburg, Russia and Eastern Kentucky, USA. These areas were selected for two reasons. Firstly, because, in each area, educational reform has recently been introduced in an effort to raise standards, and secondly, because economic hardship and limited vocational futures appeared to undermine the life chances of a significant proportion of the young people concerned.

Within the UK, recent interest in Eastern European education has tended to focus largely upon pedagogic factors, in particular the use of whole class, interactive teaching. Close examination of mathematics teaching in Hungary by David Burghes and his colleagues at Exeter University has been influential in Government circles and has informed current initiatives in mathematics teaching such as the national numeracy hour. Since the seminal study of Bronfenbrenner (1971) surprisingly little research, however, has been undertaken to compare Eastern European children's attitudes and orientation towards education and schooling with those of Western societies. This study seeks to redress this imbalance although it should be emphasised that the geographical areas chosen are illustrative and not intended to be representative of particular nations.

The present study explored student attitudes and reported behaviours, which the (largely USA) literature suggests are significant variables in explaining educational performance. These were (a) children's perceptions of, and satisfaction with, their current performance and work rate in school and at home, (b) the importance ascribed to effort, as opposed to ability, in explaining achievement, (closely allied to this issue are peer influences, in particular, the acceptability of outward shows of educational striving) and (c) the value placed upon education as an end in itself and/or as a means to a desirable economic/vocational future. Given the extant literature, it appeared reasonable to assume that educational performance will be reduced where satisfaction with mediocre performance is widespread, notions of fixed ability as the key to success prevail, an anti-work ethic exists within peer cultures and educational success is not perceived to have significant intrinsic or extrinsic value. In order to consider the impact and importance of the above issues, it was necessary to consider broader cultural and contextual influences, which may underpin or mediate such variables. Thus, the final part of this chapter endeavours to contextualise the study findings to the particular socio-economic concerns of each of the three regions.

In summary, the study described here aims to provide greater understanding of the nature and potential influence of key variables upon motivation, and thus upon educational performance. The study has been conducted in three very different geographical areas which, somewhat paradoxically, are confronted by a shared concern, how to raise educational performance in the face of harsh economic circumstances.

2. SAMPLE

2.1 Survey

The sample for the survey consisted of 3,234 children, all of whom attained their fifteenth birthday during the academic year of the survey. In each country the 'home' university (Morehead University, Kentucky; Hertzen University, St. Petersburg; and Sunderland University) had responsibility for ensuring the representativeness of the sample for the locality under investigation.

Sunderland (n = 1286): In selecting schools representative of the city, one independent and eight comprehensive schools were approached. These were selected on the basis of their size, examination profile, geographical location, religious denomination and socio-economic mix (n.b. as none of the areas in the study has a significant proportion of children from ethnic minorities, this variable was not a feature in the selection of schools). Each school agreed to the participation of all children in Year 10 in the survey, which was conducted by members of the research team. From the outset, the co-ordinator of the Russian end of the project took part in the Sunderland survey in order to ensure consistency of approach across countries.

Russia (n = 1324): Subjects were drawn as whole class groups from 21 schools in the city of St. Petersburg and its immediate suburbs. Although Russian schools are generally comprehensive in their socio-economic intakes, a representative sample of gymnasium classes, which cater for more able scholars, was included. The survey was undertaken by a research team based at Hertzen University, St. Petersburg, led by the co-ordinator who had taken a significant role in both the questionnaire's construction and subsequent translation into Russian, and the Sunderland data collection.

Eastern Kentucky (n = 633): As Kentucky is not densely populated and the numbers of children in each school year tend to be greater than those in Sunderland or St. Petersburg, the schools are physically distant from one another. Schools in the region have broadly similar intakes and can be said to be fully comprehensive. Given the distances involved, almost every child attends his or her local high-school. For this reason, only three schools were selected for the present study. Each of these performed close to the State mean in a series of 1996 externally produced performance measures, which function in a way similar to SATS in England. Children from the two large cities in the State, Louisville and Lexington were not included in the study as the research team wanted to concentrate upon rural Eastern Kentucky where economic decline mirrored that of north-east England and where there is evidence of a conflict between the region's traditional values and those messages provided by teachers and educationalists (Saragiani *et al.,* 1990; Wilson *et al.,* 1997).

All children in the appropriate age-group in each of the schools took part in the survey, which was conducted by two members of the Sunderland team assisted by partners from Morehead University, Kentucky.

2.2   Individual interview

The following year, 130 children aged 14 and 15, from schools which took part in the survey, were individually interviewed at length about issues covered in the questionnaire. Students were selected randomly although a quota system, based on ability, was utilised to ensure that the full ability range was represented.

Children in Sunderland and Kentucky tend to perform at the lower end of their national cohorts in national tests (such as GCSE). Although there are no similar data which permit analysis of the relative performance of St. Petersburg children within a wider Russian population, its status as a major educational and cultural centre suggests that the children in the sample would be relatively high performers in national terms. Thus, not only do Russian children appear to be generally higher performers than those in England and the United States but the samples would seem to draw upon cohorts performing above (St. Petersburg) and below (Sunderland/Kentucky) their own national levels.

3.   METHOD

3.1   Survey

All the students in each class group, who were present on the day of the survey, completed a detailed questionnaire which explored the following issues:
1. attitudes to school, teachers and school subjects,
2. attributions for success/failure
3. behaviour at home and in school, and
4. expectations/wishes for the future.

In introducing the questionnaire to the children it was pointed out that the researchers were interested in commonalities and differences in children from the three countries. Following the valuable guidance of Assor and Connell (1992) the children were asked to become valued reporters and it was stressed that the team desired their perceptions (rather than 'veridical assessments' which may be held by others). Thus there were no right/wrong or desirable/undesirable answers. In order to minimise the potential influence of national chauvinism [a charge which may be levelled at some of the large-scale comparative studies (Wainer, 1993)] it was stressed that the goal of the investigation was to get greater insight into the contemporary lives and attitudes of teenagers; thus their responses should not be seen as placing their country in a favourable or unfavourable light.

The items employed were derived from a wide literature on student achievement, with several questions similar to those used in a survey of student attitudes produced for the UK National Commission on Education (Keys and Fernandes, 1993). The format of the questions differed somewhat, however, and rather than circling a number on a numerical scale, as was employed in the NCE study, most of the questions in the present survey required the students to circle one statement from a selection of four or five in ranged order (question type *a* – see appendix p.114 for examples). Other questions required the respondent to rank a series of statements (type *b*) or select from a simple yes/no alternative (type *c*). In order to minimise student confusion, numerical codes were not placed in the margins of the questionnaires and transfer of the data into numerical form took place subsequently.

Piloting of the questionnaire in the three countries indicated that the presence of a second individual in each classroom would help to ensure that the procedure would work effectively. Each question was read aloud by a member of the research team and the children were asked to enter a response on their questionnaires. This method ensured that children with reading difficulties were able to respond, although a small proportion required assistance from members of the research team. Before each completed page was turned over, the researchers scanned the responses to ensure completion. This was undertaken in a swift fashion in order that the children would not feel that attention was being paid to the content of their responses. In order to ensure anonymity, children were asked to make certain that they could not be identified by their responses. As Kentucky operates a highly inclusive approach to special education it was found that a small number of classes contained a child with severe learning difficulties for whom the survey was incomprehensible. Given that in St. Petersburg and Sunderland such children would be placed in special schools it was considered that the omission of such children from the study would not threaten representativeness. Some words in the Kentucky questionnaire were substituted to reflect subtle differences in language use (e.g. smart for clever) although in all other respects the content was identical.

The method of administration of the questionnaires resulted in response rates, which were uniformly high.

3.2   Individual interview

In each country, children aged 14 to 15 years took part in an individual interview lasting approximately 45 minutes. All interviews were tape recorded and transcribed. Anonymity was assured to the children who all agreed to take part in the interviews. Questions were designed to flesh out issues arising from the survey data and to ensure that the findings obtained were also in evidence when children engaged in more in-depth discussion (a form of triangulation). An interesting and novel aspect of the study was that the children were interviewed by researchers of other nationalities (e.g. Russian and American researchers asked essentially the same questions of Sunderland children that the English researchers asked in Kentucky and St. Petersburg.) The use of 'foreign' interviewers permitted the posing of 'naïve' questions and reduced any tendency on the child's part to assume that the researcher already understood existing contexts and current practices.

4.   RESULTS

4.1   Perceptions of, and satisfaction with, current performance and work rate

Findings that American children and their parents were more easily satisfied with average or low performance than those of south-east Asia (Stevenson and Stigler, 1992) were reflected in the present comparison of Russian and Western contexts. In their responses to a number of items, both English and American samples consistently demonstrated strikingly higher levels of satisfaction with their educational achievements and their work rate in school than did their Russian peers. Table 5.1 indicates responses to three questions concerning the child's perception of his/her schoolwork ability and what they consider to be the views of their parents and teachers.

Table 5.1. Child's perception of self, parents' and teachers' views of his/her schoolwork ability (%)

	St Petersburg			Kentucky			Sunderland		
	Self	Teacher	Parent	Self	Teacher	Parent	Self	Teacher	Parent
Very good	2	3	4	14	16	34	11	12	23
Good	21	18	19	35	37	35	51	46	47
Average	46	42	32	42	34	23	33	35	23
Not very good	26	28	33	6	8	5	4	5	5
Poor	5	9	12	3	5	3	1	2	2

Chi-square analysis of self: X^2= 596, df=8, p<.000005.
Chi-square analysis of teacher: X^2= 584, df=8, p<.000005.
Chi-square analysis of parent: X^2= 931, df=8, p<.000005.

These figures suggest that children in Sunderland and Eastern Kentucky have a far higher perception of their own abilities than their Russian peers, a view not borne out by international comparative studies or more anecdotal school observation in the three countries.

Children's opinions of their teachers' views mirror their self-perceptions. When children were asked to indicate what they believe their parents think, however, the discrepancy is striking. In the opinions of the respondents, Sunderland parents are considered twice as likely as their children to consider the respondents to be very good with few seeing their children as poor. The Kentucky ratio of parental to child 'very goods' (5:2) is even higher, with over a third of parents considered to view their children in the most positive category. In contrast, the perceptions of the St. Petersburg sample appear to be in an opposing direction.

To what extent does this picture of satisfaction manifest itself in other ways? Table 5.2 summarises responses to a number of key questions.

Table 5.2. Children replying affirmatively to survey questions (%)

	St Petersburg	Eastern Kentucky	Sunderland	Chi-square	Sig. Level
Are you satisfied with your school achievements?	27	63	59	$X^2=353$, df=2	P<.0001
Do you usually work as hard as you can in class?	39	64	66	$X^2=220$, df=2	P<.0001
Do you usually work as hard as you can on your homework?	51	48	42	$X^2=21$, df=2	P<.0001
Could you improve your performance a lot?	91	75	72	$X^2=163$, df=2	P<.0001

When asked whether they usually worked as hard as they could in class, English and American children tended to answer affirmatively; an approximately similar proportion of Russian children said that they did not. Interestingly, there was less difference in the proportions replying that they usually worked as hard as they could on their homework. The time actually spent on homework, however, was markedly different, with Russians spending significantly more time both on weekdays and weekends on homework tasks. A puzzling, seemingly unrelated, finding was that Russian children appeared to go to bed much later than the other two groups on week nights but not on weekends. Subsequent investigation indicated that this was because the heavy homework demands were not substituted for the usual social and recreational activities of adolescents in USA/UK but, rather, were conducted in addition to normal teenage pastimes.

4.2   The importance of effort

In the present study, children in the three countries were asked to rank in order four reasons for succeeding in schoolwork. These were (a) working hard, (b) luck, (c) being clever and (d) being liked by the teacher. Responses are presented in Table 5.3.

Table 5.3. Ranked responses to the question 'What is the most important reasons for success in schoolwork?' (%)

	St Petersburg				Eastern Kentucky				Sunderland			
	1st	2nd	3rd	4th	1st	2nd	3rd	4th	1st	2nd	3rd	4th
Luck	6	8	24	62	10	18	44	28	10	13	50	27
Being clever	49	35	11	5	16	54	22	8	11	65	19	5
Liked by teachers	10	22	47	21	5	11	26	58	3	6	26	65
Working hard	36	35	17	12	72	15	8	5	80	16	3	1

Chi-square analysis of luck: $X^2 = 403$, df=6, p<.000005.
Chi-square analysis of being clever: $X^2 = 540$, df=6, p<.000005.
Chi-square analysis of being liked by teachers: $X^2 = 582$, at df=6, p<.000005.
Chi-square analysis of working hard: $X^2 = 610$, at df=6, p<.000005.

The findings show a marked difference between groups, with the Russian children displaying significantly greater emphasis upon ability as the key factor in achievement. The great majority of American and English children, in contrast, see effort as the major determinant. The interview transcripts, however, indicated that this issue is complex. Although the Russian students recognised the importance of ability:

I think it depends on whether you are oriented, predisposed to (academic) study … I have a friend, who is not predisposed to one subject – so he can only get 'three' [on a five-point scale where 5 is the 'best' score] there.

A lot of people said to me – if a person is not 'brain-prepared' to this particular area – he can spend hours and hours – this won't help.

I think there are certain subjects which require some special ability or talent. For me it is Physics. Hard work here may not necessarily bring best results.

There was a more general view that effort will be productive of creditable performance, whatever one's ability:

It is possible [for a person who is not clever to do well at school] if he tries really hard, sometimes learning things by heart.

I don't think he can get the same result as the person who is more clever. But he may get a significant success.

… if there is no brain, hard work alone can't get you excellent results. Though, I think one could get all 'fours' [on a five-point scale] still.

It was widely recognised, however, that abilities are necessary for the highest achievements:

If the person who is not very talented sees the task and thinks – it is very difficult for me – he'll come home and work hard for the whole day. If he comes to school next day he will get a good mark. And if a very talented person works hard, only in this case will he get an excellent mark.

[I am] not sure that everything can be achieved with hard work.

… with most [pupils] it is true, they cannot do well if they are not clever.

You can get a good result learning things by heart and sitting and sitting, but it is not worth it. Brainpower will help more.

… if you just learn something by heart without understanding – it is silly – no use. I used to know a girl who knew the whole chemistry textbook by heart – but she didn't understand a word there.

A similar pattern, though with some different overtones, can be discerned from the USA and UK interviews. In general, making sustained effort is seen as the road to success in both school and life:

Anybody can get real good grades if they work hard – if you work hard and study and do all your work. (Kentucky)

Somebody can be really clever, but not put the work in and they wouldn't get very far, where someone who worked hard would achieve the grades. (Sunderland)

If you work hard enough, where you want to go, you can do it in the end. (Sunderland)

If you work hard you can reach your goals if you work hard at them. So, like, if your goal is to be in Harvard, you can work toward your goal and get into Harvard. (Kentucky)

As long as you're a hard worker, you're gonna go far in life. (Kentucky)

I think that people who can't do things, don't give it a chance, I think if they are trying, they will eventually start learning. (Sunderland)

You have to work hard. I mean clever is good, but working hard will beat clever. (Kentucky)

The US students were a little more positive about the level and certainty of success due to effort, whilst the UK students were more tentative, but the difference was not great.   Equally, many in both groups saw effort as entitling those who made it to some recognition of moral worth, in its own right, independently of the level of success:

> The way I look at it, as long as you try, then you've tried, that's all you can do, to the best of your ability. (Kentucky)

> As long as you do your best then you have got what you want because you have done your best.   And that is all you can achieve, your best. (Sunderland)

It seemed that many teachers recognised and rewarded effort as a virtue in itself.

> If you don't work very hard, if you don't get it done, you won't get as good grades.   I think [teachers] count effort more here than anything. They just want you to try your best, and they give you a good grade just for trying as hard as you can. (Kentucky)

Certainly pupils saw the rewarding of effort as morally compensating for differences in innate abilities.

> Not all people are born ya know, smart and, uh, as long as you work your hardest, it's all, that's all you can do. (Kentucky)

> Not everyone can be really clever, but as long as you can do your best people are going to be proud of what you have done. (Sunderland)

> If you're not clever, you are never going to be clever but at least you can try your hardest 'cause, like, clever people it comes natural to them, so they're not really making any achievement by just doing what naturally comes to them. (Sunderland)

Some students, in both the USA and the UK, saw making sustained effort as a means of increasing ability.

> ... and, you know, maybe, the way you worked on it; maybe that can help you later, you know, on how to work on something else hard. (Kentucky)

> If you work hard, you can achieve cleverness, I guess.   So, if you just keep workin' hard, you'll ... you'll just probably be pretty smart, when you just work hard and work hard. (Kentucky)

I think that if you work hard you are getting brainier, you are getting cleverer. (Sunderland)

I think working hard because if you're very clever then if you work hard you might get even more clever. (Sunderland)

However, in both groups, as with the Russian students, there was a recognition that levels of talent, or intelligence, made learning easier for some and harder for others at the extremes.

Some kids are kinda born a little bit smarter. Things come a little bit easier to them, you know. (Kentucky)

Sometimes people seem to do well and they don't really work. (Sunderland)

Like if you are quite daft, you wouldn't be able to do the level of English work, if you are quite brainy you would be able to do the work. (Sunderland)

When you're clever, you just kinda have it, but almost anybody could work really hard and still get good grades. (Kentucky)

And, as with the case of the Russian student who knew the chemistry book, effort might not always be well-applied.

If you are clever you know what to write and you know what to do, but if you are just working hard it doesn't prove that you are being clever, you could be just copying down something and anyone can do that. (Sunderland)

There was also a recognition that in some subjects, special talent may be necessary to achieve the highest levels.

In some classes if you work really hard you'll get a good grade and you'll learn something. In some classes [music] you have to be just really smart, because you can't just work really hard and get a good grade, you have to know it completely. (Kentucky)

And one student expressed the complexity of the relation between effort and ability as it obtains in all three countries.

I think that not bright people can still work hard and get the results, but I think you still have to have a bit of cleverness in you and talent. (Sunderland)

It is important to recognise, however, that attributions to effort and recognition that one can't succeed without working hard do not necessarily result in high work rates. An important influence on the commission of effort is the acceptability to peers of overt striving to achieve within the adolescent subculture. Findings from the present study reinforce the growing notion that an anti-work climate exists in many English classrooms where behaviour may be poorer than in many other industrialised nations. Table 5.4 indicates that poor behaviour, on the part of peers, was more evident in the experience of the Sunderland and Kentucky samples and this appeared to impact negatively upon respondents' work rate. Of those who reported being influenced by their classmates with regard to schoolwork, three times as many St. Petersburg children indicated that peers made them work harder, rather than less hard. In contrast, the Sunderland children indicated a ratio of almost four to one, in the negative direction. Kentucky children indicated a relatively even split between positive and negative effects.

Table 5.4. Responses to the statement, 'Children in my classes do not always pay attention and lessons are disrupted by bad behaviour'. (n.b. respondents were asked to select from one of four options)

Response Options	Region					
	St Petersburg		Kentucky		Sunderland	
	No.	%	No.	%	No.	%
'Often'	102	7.7	311	49.3	475	37.1
'Sometimes'	454	34.3	257	40.7	678	53.0
'Rarely'	319	24.1	46	7.3	107	8.4
'Almost never'	447	33.8	17	2.7	19	1.5

Chi-square analysis: $X^2=1067$, df=6, p<.000005.

4.3 The purpose and value of education and of being educated

In examining a range of behaviours in the classroom and outside of school, the present study found a consistent pattern of response, which may help to explain the higher level of educational performance in Russia. St. Petersburg children spend significantly more time on homework (29 per cent reporting that they were likely to spend more than three hours on homework per evening compared with 4.8 per cent and 4.3 per cent for Sunderland and Kentucky samples respectively). Furthermore, the Western samples expressed considerably less belief in the potential impact of homework upon educational achievement (this was particularly strong for Sunderland). In comparison with the St. Petersburg sample, Sunderland and Kentucky children watch more television, play more sports, undertake less reading and engage less regularly in cultural pursuits such as visiting theatres, museums and concert halls.

Despite evidence suggesting that Russian children are becoming increasingly alienated from school (Nikandrov, 1995) the St. Petersburg sample indicated the most positive attitudes – 78 per cent 'very much' or 'quite' liking school compared with 46 per cent for both the Sunderland and Kentucky samples. Russian children also reported a leaning towards academic subjects in contrast to the other groups' preference for practical activities such as art, P.E. and drama.

Although the Sunderland sample saw qualifications as far more important than the Russian/American samples in getting a job (twice as many ranking this as the most important factor), their views as to their vocational future appear somewhat optimistic given published employment figures for the region. Table 5.5 reflects the children's perceptions of what they thought they would be doing at age 18.

Table 5.5. Responses to the question, 'What are you most likely to be doing when you are 18?' (%)

	St Petersburg	Kentucky	Sunderland
Have a job	16	21	30
FE/Tech College	13	21	32
University	69	53	29
Employment training	0	1	6
Stay at home and do nothing	0	3	1

Chi-square analysis: $X^2= 514$, df=8, p<.000005.

Interestingly, a much smaller proportion of the Sunderland sample anticipated continuing their education into adulthood.

When asked to choose from a list of six options why they might wish to work hard on schoolwork, it is not surprising, given the perceived link noted above, that Sunderland and Kentucky teenagers prioritised the need to get qualifications (75 per cent and 59 per cent of the respective samples selected this first) while the St. Petersburg sample (53 per cent) stressed a desire to be an 'educated person'. Although, in interview, almost all of the Russian adolescents offered the view that education was important for their future careers, a significant number (approximately one third) stressed a strong belief in the importance of being an educated person.

I think education is important in itself. Even if I don't get an important job, it is good to be intelligent.

It is also important for your personal development – the less you know – the harder it is for you to live.

... there can't be any useless knowledge at all.

I like to have knowledge myself, and I also understand I'll need it in future.

I just like attending classes and getting some new knowledge.

I can't stop study. (interviewer) – Do you think it is because of your personality? – Certainly.

It gives you choice, it also helps you to see your potential and to use in a proper way.

I can't find a proper word – it is like self-realisation ...

I think, we are to become persons, but not machines for making money. A man doesn't live for money only ... something else is also important – spiritual values, a soul ... not only material things.

I think you still need education, well, for yourself, to know something – to study history, to know your country better, to have some vision of the world.

[Being educated] ... is more than important. It may be the aim of life.

Such statements were only faintly and occasionally mirrored in the USA and UK interviews. For a couple of Kentucky students, education seemed a means of social standing.

I just want to know, know about everything. Just know a little bit about everything, so if anybody asks me any questions, or anything, I'll know what I'm talking about. (Kentucky)

You don't want to go in a room and be like the only one that doesn't, really know anything about the subject, (Kentucky)

whilst, for another:

It kinda broadens your horizons ... kinda, gives you a different view of different things, (Kentucky)

whilst another thought being widely knowledgeable would be:

Pretty cool, 'cause you could, like, do all kinds of stuff; you could help out the community and things like that. (Kentucky)

But another small minority was more ambivalent about the role of education in their community.

> I think it's pretty important just so you ... do know about, like history and stuff like that. I don't think it's of major importance, but yeah, I do think it's important. (Kentucky)

And another said:

> I guess, it's important to be educated. I don't know. I don't really know anybody like that. (Kentucky)

For another minority, the value of education beyond basic skills seemed very doubtful.

> People respect you for what you've tried to learn, and stuff like that, but I really don't think that you really have to know all that stuff, 'cause 90% of the people in our society don't have a clue to anything like that. (Kentucky)

> The way people act around here, they don't think it's probably real important. To be able to talk about Shakespeare and all that stuff. (Kentucky)

> My family is from mountain country, like, and ... it doesn't matter to them, I don't think ... Shakespeare and all that don't matter to them. I mean like math, spelling, and stuff like ... is important. (Kentucky)

Amongst Sunderland students, only two expressed any view which could be construed as at all valuing education intrinsically.

> Because I want to learn how to do ... because I don't understand some things, I want to know how to do it, then I might understand more, like better, how to do things. (Sunderland)

> I want to be knowledgeable. I think you do get more respect. (Sunderland)

Almost all the students in both the USA and the UK valued education as a 'ticket' to a job, an income and a life-style.

> You probably won't get nowhere quittin' school. 'Cause, you don't have an education and if you don't have an education, it's kinda hard to get a job doing something where you can make a lot of money. (Kentucky)

If you get a good education you get good exam results at the end of it. Then you can go and get a good job, a well-qualified job, that you can get paid more for and you can lead a healthier life. (Sunderland)

I would like to be an educated person, because I want to get a good job, money. (Sunderland)

There were differences of emphasis between Kentucky and Sunderland. Sunderland 14 to 15 year-olds seemed to be more dominated by the need to get the widest general education, as defined by the English National Curriculum, in order to compete in the job market. A few Sunderland students (about 10 per cent of those interviewed, all male) aspire to fame and fortune as members of rock-bands or as professional soccer players. They devote considerable time to practise towards those ambitions, but even they are realistic about the need for qualifications as a fail-safe.

If I don't get to be a guitarist, I'll need to find something to do, like a job, and that, and if I haven't got good qualifications, I won't be able to get anything that pays lots of money. (Sunderland)

You can get more chance of getting a job if you have got education, if something happens to you in football (soccer) ... (Sunderland)

All Sunderland students were more concerned with the number and level of passes in the 16 plus GCSE examinations, than were their American counterparts in attaining a spread of good grades.

I would say that education was important, because I think, some people think that they will just get a job when they are older, like, without getting any qualifications, but now more than ever I think you do need to have some qualifications to get a job. (Sunderland)

You cannot get a good job unless you go to school and go through the education and get good grades and exams and go to University and things. (Sunderland)

Of course, the US 15 year-olds were at the start of their Senior High schooling and the critical results for employment and passage to higher education were still nearly three years away. The Kentucky students seemed a little more relaxed.

I don't want to be an over-achiever too. Like some students I know, they spend all their time on schoolwork. They don't have lives outside of school, you know. Because all they do is study. And, even my parents say that's not what you need to do. Sure you need to concentrate a lot on school, but you can't make that your whole life, or you won't have any fun. (Kentucky)

To me it's important to be smart – book smart and common sense smart. Some people are book smart and have no common sense. (Kentucky)

You gotta know the basic skills of life, too. You've got to know how to take care of yourself, and smart ain't always going to get you … You need to know how to get money to go places, to be smart. (Kentucky)

However, both groups of students saw serious risks arising from failure to strive through schooling. There was the possibility of a dead-end job at best,

Just, the other day one of my friend's sisters, she dropped out of college, and my friend's mom, she was talking about how dumb that was, because when she gets on down the road, – it's going to be, when she's wanting to get a better job, and she doesn't have a college education – it's going to really mess her up. (Kentucky)

About all you can do with just a high-school diploma is flip burgers at McDonalds. (Kentucky)

Most people, when they leave school find it a shock, to see that they won't be able to get a job, as they haven't worked at school and things. (Sunderland)

and more dramatic fates, though no doubt grounded in the life of the community, were forecast by some.

You're never going to get a job, and you're going to be poor, and you'll probably end up on welfare, and you'll probably end up pregnant when you're 16 or whatever, if you quit school. (Kentucky)

If you have no education you end up turning to being a thief because of boredom and you get nowhere in life … (Sunderland)

Both groups saw the need to defer present gratification for future gain.

What you learn will help you in the long run and stuff. Like people are already asking how this is going to help you in life, and I know that one day you'll probably use it. So, you might as well just go ahead and learn it, whether you like it or not. (Kentucky)

If you have a, you know, a real life goal that you want to do something in college, then you need to start, you know, at a young age and, you know, you need to build yourself up each year and you know, not settle for bad grades. (Kentucky)

I want to make sure that I get a good job when I leave school, so I work hard now, so I can get the qualifications I need, so I have a chance when I get older. (Sunderland)

And both groups recognised that the world was changing in ways which placed a higher value on at least some aspects of education. For the Sunderland students this seemed to have become largely focused on getting a good general education, particularly in English and maths, and going on to further or higher education, without necessarily having a set career plan. The Kentucky students emphasised particular aspects of this change.

The world is changing around. And it's going into technology, and if you don't know a lot about technology, I don't think that you're gonna know enough to get a decent job. (Kentucky)

Everythin's gonna go to computers and if ah, you don't know much about 'em, you're gonna be sorta lost. (Kentucky)

You can't get a job like my dad's any more, really. No. Only way, you got to work at McDonald's or something if you don't finish high school or college. (Kentucky)

There ain't very many jobs left that you wouldn't need education ... I don't think there's any, maybe diggin' a ditch or somethin', diggin a hole. Be about it. (Kentucky)

5.   DISCUSSION

5.1   Perceptions of, and satisfaction with, current performance and work rate

The Sunderland finding of high levels of satisfaction with present performance mirrors that of a survey of children, one year younger, in England and Wales (Keys and Fernandes, 1993). More recently, in analysing the performance of a sub-sample of eleven countries taken from the recent TIMSS study, Keys *et al.* (1997) also found that English Year 9 students had higher self-perceptions than any other country. Perhaps most striking was the finding that 93 per cent of the English students agreed, or strongly agreed, that they were doing well in mathematics compared with 57 per cent of children from Singapore, the highest performing country in the TIMSS study. While there may be cultural factors which require an expression of modesty from Asian children, this does not explain the English scores in comparison with European countries such as France (68 per cent) and Germany (69 per cent) whose performance in mathematics was also superior (Beaton *et al.*, 1996b).

At first, the picture of self-satisfaction in the Sunderland and Kentucky samples, which emerges from the data, is somewhat puzzling. While the disparity between the English and Russian perceptions could be explained by the difference between the somewhat nebulous system of reporting National Curriculum levels in England and the operation of a clearly understood regular system of grading to signal progress in Russia, this appears an insufficient explanation given the existence of a grading system in Kentucky which mirrors the Russian model. There is no reason why either Russian or American grades should be misunderstood. Indeed, St. Petersburg and Kentucky children were equally likely to respond that their parents had a good understanding of their relative performance in schoolwork (74 per cent St. Petersburg, 72 per cent Kentucky, 62 per cent Sunderland).

It seems likely that the above discrepancy is explained, not by a failure to understand absolute performance levels but, rather, by differences in what is considered desirable. In their comparison of American and Asians, Stevenson and Sigler (1992) found that although both samples of mothers were able to evaluate their children's relative academic status in relation to their peers, the kind of interpretation placed upon performance was significant. A position among peers described as 'average' by Japanese and Chinese mothers was considered 'above average' by American mothers. Furthermore, as children moved through the primary school, Asian parents appeared to require increasingly higher levels of achievement from their children in order to be satisfied, whereas American mothers reacted in the opposite fashion. Similarly, the data in the present study, reflect the (anecdotal) perceptions of Russian visitors to the UK who have often remarked upon the more positive constructions English teachers place upon children's work.

In contrast, the self-assessments of the Russian children appear to reflect the messages they receive from their teachers where the main method of motivating children to perform better is the adoption of a critical, albeit supportive, attitude to performance. Russian teachers, therefore, are more likely to point out to a child what is wrong about a piece of work than what is good. More recently, however, a greater emphasis upon self-esteem, influenced by growing awareness of Western psychology, has led some Russian teachers to question traditional practices (Karakovsky, 1993).

As is the case in the UK, there has been much publicity in the USA about poor educational standards. Stevenson and Stigler's (*op. cit.*) analysis suggested that this appeared not to have influenced American parents' perceptions of their own offspring:

> *Other* children might be performing poorly and *other* schools might be failing, but these were not problems that most mothers discerned in their own situation (p.118, emphases as in original).

The rosy picture of American parents in the primary school years appears to be mirrored by the American and English parents of the adolescents in the present survey. Although this was a survey of children's perceptions, the data suggest that different understandings of children's ability reported in the Asian-American study are repeated although, this time, it is Russians, rather than Asians, who appear to be more circumspect.

The finding that the St. Petersburg sample also tended to have more negative perceptions of their work rate may be explained by considering the nature of their classroom context where highly intensive interactive teaching makes huge demands upon attention and concentration. Thus, periods of disengagement may become more salient and personally problematic than in Anglo-American classrooms. Similarly, much greater homework demands were reported by the St. Petersburg sample. Although the relationship of homework to achievement is rather unclear, with conflicting findings emerging from research studies (Farrow *et al.*, 1999; Epstein, 1988; Olympia, *et al.*, 1994; Cooper *et al.*, 1998), it would appear that, where homework is meaningful and relates closely to classroom activity, increased time is associated with greater achievement (Anderson, 1986; Cooper, 1989; Keith, 1987). In an analysis of TIMSS mathematics scores for the English sample (Keys, Harris and Fernandes, 1997) a clear relationship between time spent on mathematics homework and performance was demonstrated. Clearly the influence of homework upon achievement is complex; in one recent study, for example, (Cooper *et al.*, 1998) it was found that while there was a weak relationship between homework assigned and student achievement, positive relationships were found when the amount students actually completed was considered. The data in the present study, of course, reflect this latter variable.

Daily homework in Russia, even for primary-aged children, is substantial and closely woven into classroom learning. Failure to complete, therefore, not only results in poor grades but also often leads to difficulties in coping with the next day's academic demands – important motivators to ensure homework is completed. This integration of class and home work should serve as an important guide to UK policymakers whose calls for increased homework demands has the potential of resulting in an increase in irrelevant, time-filling bolt-on activities.

It is often argued that an important means of raising standards is boosting children's confidence in their abilities and potential although recent research has challenged the view that such constructs as self-esteem and self-efficacy are important determinants of academic performance (Eaton and Dembo, 1997; Muijs, 1997). While it is generally agreed that the educational progress of many children can be hindered by self-doubt, it is also likely that unrealistic, over-optimistic perceptions of one's performance may reduce goal-seeking behaviour. Several studies (Stevenson, Chen and Uttal, 1990; Stevenson and Stigler, 1992), for example, conclude that unduly positive estimations of children's abilities and low expectations negatively impact upon American children's academic performance.

Perhaps a strong desire to succeed is more influential than self-perceptions. Eaton and Dembo (1997), for example, concluded that while the Asian Americans in their study demonstrated lower levels of self-efficacy than Caucasian Americans, fear of failure, stemming from family pressures, best explained their high levels of achievement behaviour.

5.2   The importance of effort

Although the focus upon effort demonstrated by the Sunderland and Kentucky samples matches the findings of Blatchford (1996), Lightbody *et al.* (1996), Gipps and Tunstall (1998) and TIMSS data (Keys *et al.*, 1997), they run counter to the widespread consensus that Anglo-American cultures conceive ability as the most important attribution for success and failure. Certainly, this appears to contradict the highly influential findings reported by Stevenson and Stigler (1992) which inform many contemporary American texts on the roots of underachievement (e.g. Ravitch, 1995). A number of possible explanations exist. Firstly, the ages of the children in the two studies differed. The children in the Asian/American study were fifth grade (Year 6 in England) whereas those in the present study were four years older (n.b. the children in the Gipps and Tunstall (1998) study, however, were substantially younger than in any of the other investigations). A second reason for this discrepancy may relate to differences in the questions asked. Unlike the present study, where prioritisation by means of ranking was sought, Stevenson and Sigler asked children to indicate the level of their agreement with the statement, 'The tests you take can show how much or how little natural ability you have.' The American sample tended to agree far more strongly to this proposition than the Asian sample. This question, however, is open to a variety of different interpretations and, in the opinion of the present writers, does not resolve the effort/ability issue. Agreement that the tests 'can show' the level of an individual's natural ability does not preclude a strong belief in the importance of effort in achievement. If one were to transfer the question to that area of American obsession, sporting performance, a belief that sporting achievements are likely to be won by those with natural ability would be unlikely to reduce children's efforts to become more skilled. The key difference, perhaps, lies in the intrinsic enjoyment of, and the cultural value placed upon, the activity concerned.

There is little evidence, either anecdotal or empirical, that Russian children's greater emphasis upon the importance of ability is a demotivating factor undermining their performance. Despite the fact that the Russian children tended to have lower academic self-perceptions than their Western counterparts, their responses indicated a stronger belief that they could achieve more if they expended greater effort. Such a pattern is puzzling, as a greater belief in the centrality of ability and lower self-perceptions of one's ability would not seem to be consonant with a greater emphasis upon one's capacity to achieve more through heightened effort. In Russian classrooms, however, there are strong expectations that children can and should achieve whatever their academic level. Thus, considering oneself as of limited ability does not generally result in a belief that one cannot make meaningful educational progress.

Clearly, the present study could be criticised on the grounds of its methodology. The use of a simple ranking system only permits the relative importance of constructs to be indicated. The fact that ability is chosen before effort by a greater proportion of the Russian children does not rule out the likelihood that effort is also considered to be a highly influential, indeed essential, element in achievement – indeed, the interaction of both variables was a feature of the interviews. An oversimplified, Manichean view of these variables, however, may be an important failing of other international comparative studies (Stevenson and Stigler, 1992). Furthermore, the present study was not able to consider more complex issues such as any differences children may hold in their attributions for failure as opposed to success.

Despite these weaknesses, the data obtained would appear to support findings from other studies (Blatchford, 1996; Lightbody *et al.,* 1996) that English children tend to attribute their academic performance to internal factors. The debate on international differences in performance, however, has largely focused upon differences in internal attributions (rather than between internal and external factors), in particular the relative importance of ability and effort (e.g. Gipps and Tunstall. 1998). This latter distinction, reproduced in the present study may be oversimplistic, and further analyses of motivational factors may benefit from the considerably more sophisticated psychological models which are currently available (e.g. Simpson *et al.*, 1997; Boekaerts, 1994; McInerney, 1997).

In much of the attribution literature, it is assumed that highlighting effort rather than ability as a key feature of performance will have an important impact upon levels of motivation. While this may be a valuable enterprise, there is little value in recognising the centrality of effort to high performance where peer pressures delimit the outward showing of enthusiasm or effort. Although the Russian children's emphasis upon ability might seem undesirable, the positive attitude to academic endeavour evidenced in the classrooms, and the value ascribed by teachers and children to hard work and diligence, more than compensated. Future attributional studies may need, therefore, to place far greater emphasis upon contextualising their findings to specific educational and sociocultural environments.

5.3   The purpose and value of education

The finding that Russian children largely enjoy attending school confirms other studies (e.g. Glowka, 1995). In reality, however, whether school is liked may be less important than the extent to which it is perceived to be valuable. The dislike of mathematics which has been frequently noted of Japanese children (e.g. Keys *et al.*, 1997), does not, for example, appear to result in poor performance. Indeed, it has been suggested that in order to raise their test scores, children may need to be taught in such a way that they will like the subject far less than they do at present (Atkin and Black, 1997). The complex relationship between positive attitudes to school and achieve-ment was demonstrated with the Sunderland sample where it was notable that the two schools with superior public examination records, occupied opposite ends of the 'liking of school' continuum. It is, perhaps, a recognition of the value of education, rather than the extent to which it is enjoyable, which results in that deferment of gratification (i.e. persevering with school-work in the light of more attractive alternatives) which is crucial for long-term success.

Even where education is highly valued, there may be very different underlying reasons which makes its esteem differentially susceptible to socio-economic transformation. In comparing Russia with high performing Asian countries, such as Singapore, interesting differences emerge. In both countries, high levels of academic performance reflect educational motivation but the reasons for this differ. Despite the long-standing cultural emphasis upon the importance of education in Asian countries, rapid economic growth in Singapore has resulted in education becoming an instrumental and essential means to an important economic end; without high qualifications there is usually a limited vocational future. In contrast,

Russians have tended to perceive education as an end in itself; to be articulate, literate and cultured has traditionally been highly valued by Russian society and academic and intellectual ability is greatly prized.

Russian education does not exist in a vacuum and it is unclear to what extent rapid economic and social upheaval will undermine the historical value which has been accorded to the intellectual and the academic. Although a significantly high proportion of the Russian children in the present study reported that they enjoyed school, Russian educationalists are beginning to express concern about declining motivation and rising drop-out rates. Interestingly, the ability to gain wealth rapidly by means of a range of entrepreneurial activities for which qualifications are irrelevant appears to be encouraging a form of educational bifurcation: greater numbers of children appear to be becoming disenchanted with school while, at the same time, applications to university courses which offer the promise of economic prosperity are rising (Kitaev, 1994).

A strong historical emphasis upon the value of scholarship is less evident in America, and Sedlak *et al.* (1986) point to a lack of intellectual commitment stretching back to the 1920s. Coleman (1961) similarly highlighted the anti-intellectualism of American high-schools where social and sporting success have often represented the pinnacle of achievement. Interestingly, however, these values found their source not in adolescent subculture but, rather, within the wider adult community. This is less true of Asian American children, however, who, greatly influenced by parental demands for academic success (Siu, 1992), appear to be less prepared to embrace anti-intellectual high school values and continue to outperform their peers academically.

In their study of anti-intellectualism in US schooling, Howley *et al.* (1995) argue that the instrumental worth of education is not, as in Asian countries, balanced by a recognition of its intrinsic value in the intellectual development of the individual but, rather, is merely considered to be a key to a 'good job'. This, together with strong social pressures to conform to anti-intellectual high-school values may help to explain the perception of widespread underachievement in the United States.

In considering American high-schools, Tye (1985) lamented a growing instrumentalism, which militated against the intrinsic valuing of education.

> The belief that the reason a person goes to school is to get a good job and earn more money as an adult has robbed our society of two important values. First of all, it deprives young people of the feeling that what they are doing now is important. All the rewards seem to be somewhere in the future. Secondly, it deprives society of the understanding that learning has value in itself and not just as a saleable commodity. This greatly

reduces the range of knowledge that is considered worth having, and creates a population of narrowly-educated citizens (pp. 337-338).

In many Western adolescent subcultures where the deferment of gratification runs counter to contemporary values ('I want it; and I want it now!'), even the promise of later employment may be insufficient to motivate many teenagers to invest in the process of education. In communities such as Sunderland where even the 'carrot' of paid employment may often be considered unattainable, it is hardly surprising that motivation to study is frequently low.

Where the value of education is widely perceived in instrumental terms (i.e. as a means of gaining better employment) an important motivational factor is likely to be the extent to which one truly believes that there is a route from disadvantage. In a study of young adults in four geographical areas of Britain, Ashton and Maguire (1986) found relatively high levels of hopelessness in Sunderland. Here, rather than being depressed by the experience of unemployment,

> ... many respondents appeared to face the prospect of long-term unemployment with resignation, acknowledging the fact that all young people in the area were susceptible, due to the shortage of jobs. By comparing their situation with that of their peers in Sunderland, they had much lower aspirations than respondents in other areas (p. 5).

Furthermore, there appeared to be little criticism of the performance of their schools in preparing them for the world of work, thus:

> ... reflecting a certain fatalism that there was little the schools could have done about their situation (p. 30).

Russian children, similarly faced by a bleak economic future, appear to be reacting in a very different fashion. The highly unstable economic situation in which entrepreneurial skills can bring about immense wealth has resulted in a shift from the traditional regard for education to the development of individualistic goals of success and prosperity (Nikandrov, 1995).

> Roughly speaking, the more education one has nowadays, the less money one earns (Nikandrov, *op. cit.*, p. 54).

Given the imperative to make money, many higher education subjects are no longer greatly valued and the most able students are increasingly opting to study those subjects that offer the greatest financial rewards (economics, finance, law, foreign languages). Kitaev (1994) cites a 1991 survey of 15,300 school leavers in the Moscow region which indicated that a growing number were willing to accept jobs which required little education but,

... promise more 'grey' (tip-taking and the like) income – waiter, hairstylist, taxi driver etc (p. 117).

The long-term impact of these changes upon educational motivation and academic performance cannot be deduced. Whether performance will decline, or, as in Asian countries, be seen as essential for the achievement of economic security is unclear.

6. CONCLUSIONS

The results reported in this chapter may help to explain some of the more important reasons why children in the United States and England perform relatively poorly on international measures. While it should be recognised that the regions sampled are not representative of their countries, they do serve to highlight the complex relationship between educational practice, sociocultural factors and economic development which all play a part in student motivation and achievement.

Findings from the present study report a range of attitudes and behaviours on the part of the Sunderland and Eastern Kentucky samples, which are inimical to the achievement of high educational standards. The trend is similar to recent studies comparing American and Asian students in very many respects: it differs from these, however, in the strong emphasis placed upon effort as a determinant of success by both English and American samples. The findings from this study may lead one to conclude that effort/ability attributions may be less important in predicting goal seeking behaviour than commentators have suggested. Perhaps more important are children's familial, peer and cultural perceptions about what constitutes real and meaningful educational achievement and the extent to which this is seen to be of such intrinsic or extrinsic value as to evoke significant effort.

For many children living in each of the three geographical areas studied, education appears to have a marginal role in their adult economic futures. Where Sunderland children appear to be have become fatalistic and apathetic, Russian children (at least, those in the capital cities) appear to have become more entrepreneurial, more independent and more confident in their own abilities (Kitaev, 1994; Nikandrov, 1995; Poznova, 1998).

Although Eastern Kentucky has little in common with St. Petersburg, it does share with Sunderland a strong tradition of mining and the dispiriting experience of coalfield closures. Its traditional economies, however, have been tobacco and mixed farming although, as in Sunderland, Asian industrial companies are now perceived as important sources of future prosperity. Kentucky is an area marked by relative poverty, where high standards of

education have not been greatly valued and the accusation that one is '...
goin' beyond yer raisin'' is an instantly recognisable criticism of academic
high fliers. Kentucky's response to underachievement has foreshadowed
developments in the UK. It has introduced legislation with a major emphasis
upon changing classroom practice, reforming teacher training, increasing
accountability primarily through rigorous testing and stressing the
importance of parental involvement (particularly at the pre-school level).
While the performance of schools is now discussed and debated, and there is
evidence that test scores have increased (Olson, 1994), there are few signs
that education has become truly valued within the wider communities.

Barber (1996) reports the findings of a study of the attitudes towards
education of 30,000 English adolescents undertaken during the 1990s.
Subsequently, under a heading, 'Can this be true?' (p. 83), he reflects upon
a picture of wide-scale student alienation, apathy and disaffection. A
tendency to negativity is also a feature of the present study, and very similar
findings to those reported in this chapter been replicated in a more recent,
large-scale study (Elliott *et al.,* submitted) of nine and ten year-olds from the
same regions as their older peers. While many bemoan the apparent shift
from enthusiastic youngster to apathetic teenager, the findings from our
study of primary school children suggest that cultural differences in attitudes
towards education are well-established by the age of nine years. Thus, the
old Jesuit maxim, 'Give me a child at age seven ...' may apply to key
elements of children's educational motivation and ultimate attainment.

The findings presented in the present paper offer no clear answer to the
question 'What matters?' They do suggest, however, that the higher
attainment of many poorly resourced (Chuprov and Zubok, 1997), Eastern
countries may be closely tied to motivational factors which will not be
addressed merely by tinkering with classroom organisation and instruction.
While it appears highly likely that international differences in the dynamics
of school and classroom help to explain varying performance (Hufton and
Elliott, 2000), these involve much broader processes than has been generally
recognised by many education reformers. In England, recent Government
attempts to raise standards have largely involved the handing down of
prescribed teaching and learning strategies underpinned by a strong 'back to
basics' orientation. It is to be hoped that in one's attempts to raise
educational performance in the USA, an over-emphasis upon discrete
aspects of pedagogy will not lead to a neglect of important linkages within
the totality of classroom practice and with broader sociocultural factors,
many of which, however, are not easily situated within Western cultures.

REFERENCES

Anderson, B. (1986) *Homework: what do national assessment results tell us?* (Princeton, N.J.: Educational Testing Service)

Ashton, D.N. and Maguire, M. (1986) *Young Adults in the Labour Market. Research Paper 55* (London,:Department of Employment)

Assor, A. and Connell, J. P. (1992) The validity of students' self-reports as measures of performance affecting self-appraisals, in D. H. Schunk and J. L. Meece (eds.) *Student Perceptions in the Classroom* (Hillsdale, New Jersey: Lawrence Erlbaum)

Atkin, J.M. and Black, P. J. (1997) Policy perils of international comparisons: the TIMSS case, *Phi Delta Kappan*, pp. 22-28 (September)

Barber, M. (1996) *The Learning Game* (London: Victor Gollancz)

Beaton, A. E., Martin, M. O., Mullis, I. V. S, Gonzalez, E. J., Smith, T. A. and Kelly, D. L. (1996a) *Science Achievement in the Middle School Years: I.E.A.'s Third International Mathematics and Science Study (TIMSS)* (Chestnut Hill, MA: Boston College)

Beaton, A.E., Mullis, I. V. S, Martin, M.O., Gonzalez, E.J., Kelly, D.L. and Smith, T.A. (1996b) *Mathematics Achievement in the Middle School Years: I.E.A.'s Third International Mathematics and Science Study (TIMSS)* (Chestnut Hill, MA: Boston College)

Bishop, J.H. (1989) Why the apathy in American high schools? *Educational Researcher*, 18, pp.7-10.

Blatchford, P. (1996) Pupils' views on school work and school from 7 to 16 years, *Research Papers in Education*, 11(3), pp. 263-288.

Boekaerts, M. (1994) *Motivation in Education: the 14th Vernon-Wall Lecture* (Leicester: British Psychological Society).

Bridgman, A (1994) Teaching the old bear new tricks, *The American School Board Journal*, 181(2), pp. 26-31

Bronfenbrenner, U. (1971) *The Two Worlds of Childhood* (London: Allen and Unwin)

Burghes, D. (1996). Education across the world: the Kassel Project. Paper presented at conference 'Lessons from Abroad: Learning from international experience in education' (London)

Burton, D. (1997) The myth of 'expertness': cultural and pedagogical obstacles to restructuring East European curricula, *British Journal of In-service Education*, 23(2), pp. 219-229

Chuprov, V. and Zubok, I. (1997). Social conflict in the sphere of the education of youth. *Education in Russia, the Independent States and Eastern Europe*, 15(2), pp. 47-58

Coleman, J. (1961) *Adolescent Society: The Social Life of the Teenager and its Impact on Education* (New York, Free Press)

Cooper, H. (1989) *Homework* (White Plains, N.Y.: Longman)

Cooper, H., Lindsey, J. L., Nye, B. and Greathouse, S. (1998) Relationships among attitudes about homework, amount of homework assigned and completed, and student achievement, *Journal of Educational Psychology*, 90(1), 70-83

Covington, M. V. (1992) *Making the Grade: A Self-Worth Perspective on Motivation and School Reform* (Cambridge: Cambridge University Press).

Dweck, C. (1986) Motivational processes affecting learning, *American Psychologist*, 41, pp. 1040-1048

Eaton, M.J. and Dembo, M.H. (1997) Differences in the motivational beliefs of Asian American and Non-Asian students, *Journal of Educational Psychology*, 89(3), pp. 433-440

Elliott, J., Hufton, N., Illushin, L. and Lauchlan, F. Motivation in the junior years: international perspectives on children's attitudes, expectations and behaviour and their relationship to educatinal achievement (submitted)

Epstein, L.L. (1988) *Homework Practices, Achievements and Behaviors of Elementary School Students* (Baltimore MD: Center for Research on Elementary and Middle Schools)

Farrow, S., Tymms, P. and Henderson, B. (1999). Homework and attainment in primary schools. *British Educational Research Journal*, 25(3), pp. 323-341

Foxman, D. (1992) *Learning Mathematics and Science (The Second International Assessment of Educational Progress in England)* (Slough, NFER)

Gipps, C. (1996) The paradox of the Pacific Rim learner. *Times Educational Supplement*, 20th December, p. 13

Gipps, C. and Tunstall, P. (1998) Effort, ability and the teacher: young children's explanations for success and failure, *Oxford Review of Education*, 24 (2), pp. 149-165

Glowka, D. (1995) *Schulen und unterricht im vergleich. Rusland/Deutschland (Schools and teaching in comparison: Russia and Germany)* (New York: Waxmann Verlag)

Hirschman, C. and Wong, M.G. (1986) The extraordinary educational attainment of Asian Americans: A search for historical evidence and explanations, *Social Forces*, 65, pp. 1-27

Howley, C. B., Howley, A. and Pendarvis, E. D. (1995) *Out of Our Minds: Anti-Intellectualism and Talent Development in American Schooling* (New York: Teachers College Press).

Hufton, N. and Elliott, J. (2000) Motivation to learn: the pedagogical nexus in the Russian school: some implications for transnational research and policy borrowing, *Educational Studies*, 26, pp. 115-136

Keith, T. Z. (1986) Children and homework, in: A. Thomas and J. Grimes (Eds) *Children's needs: psychological perspectives* (Washington, DC: NASP Publications).

Karakovsky, V.A. (1993) The school in Russia today and tomorrow, *Compare*, 23(3), pp. 277-288

Keys, W. and Fernandes, C. (1993) *What do students think about school?* (Slough: NFER)

Keys, W., Harris, S. and Fernandes, C. (1997) *Third International Mathematics and Science Study, First National Report (Part 2)*, (Slough: NFER)

Kitaev, I. V. (1994) Russian education in transition: transformation of labour market, attitudes of youth and changes in management of higher and lifelong education, *Oxford Review of Education*, 20(1), pp. 111-130

Lightbody, P. Siann, G., Stocks, R and Walsh, D. (1996) Motivation and attribution at secondary school: the role of gender, *Educational Studies*, 22(1), pp. 13-25

McInerney, D. M., Roche, L. A., McInerney, V. and Marsh, H. (1997) Cultural perspectives on school motivation: the relevance and application of goal theory, *American Educational Research Journal*, 34(1), pp. 207-236

Muijs, R. D. (1997) Predictors of academic achievement and academic self-concept: a longitudinal perspective, *British Journal of Educational Psychology*, 67(3), pp. 263-277

Nikandrov, N. D. (1995) Russian education after perestroika: the search for new values, *International Review of Education*, 41 (1-2), pp. 47-57

Oettingen, G., Little, T. D., Lindenberger, U. and Baltes, P. B. (1994) Causality, agency and control beliefs in East versus West Berlin children: a natural experiment on the role of context, *Journal of Personality and Social Psychology*, 66, pp. 579-595

Office for Standards in Education (1993) *Access and Achievement in Urban Education* (London: HMSO)

Ogbu, J. (1983) Minority status and schooling in plural societies, *Comparative Education Review*, 27, pp. 168-190

Olson, L. (1994) Dramatic rise in Kentucky test scores linked to reforms. *Education Week*, 13, p.13

Olympia, D. E., Sheridan, S.M. and Jenson, W. (1994) Homework: a natural means of home-school collaboration, *School Psychology Quarterly*, 9 (1), pp. 0-80

Osborn, M. (1997) Children's experience of schooling in England and France: some lessons from a comparative study, *Education Review*, 11 (1), pp. 46-52

Poznova, G. (1998) Russian teenagers: what do they strive for today? *Education in Russia, the Independent States and Eastern Europe*, 19 (1), pp. 19-25

Prais, S. (1997) *School-readiness, whole class teaching and pupils' mathematical attainments* (London: National Institute of Economic and Social Research)

Prelovskaya, I. (1994) New problems in the market economy, *Izvestia*, 26th June, pp. 1-2

Reynolds, D. and Farrell, S. (1996) *Worlds Apart? A Review of International Surveys of Educational Achievement Involving England* (London: HMSO)

Robinson, P. (1997) *Literacy, Numeracy and Economic Performance* (London: Centre for Economic Performance)

Rosenbaum, J. E. (1989) What if good jobs depended on good grades? *American Educator* (Winter), pp. 1-15, 40-42

Sarigiani, P. A., Wilson, J. L., Petersen, A. C. and Vicary, J. R. (1990) Self-image and educational plans of adolescents from two contrasting communities, *Journal of Early Adolescence*, 10 (1), pp. 37-55

Sedlak, M., Wheeler, C. W., Pullin, D. C. and Cusick, P. A. (1986) *Selling students short: Classroom bargains and academic reform in the American high school*, (New York: Teachers College Press)

Simpson, S. M., Licht, B. G., Wagner, R. K. and Stader, S. R. (1996) Organisation of children's academic ability-related self-perceptions, *Journal of Educational Psychology*, 88 (3), pp. 387-396

Siu, S. F. (1992) *Toward an understanding of Chinese American educational achievement (Report No. 2)* (Washington DC: US Department of Health and Human Services, Center on Families, Communities, Schools and Children's Learning)

Steinberg, L. (1996) *Beyond the Classroom: Why school reform has failed and what parents need to do* (New York: Touchstone Books)

Steinberg, L., Dornbusch, S. M. and Brown, B. B. (1992) Ethnic differences in adolescent achievement: An ecological perspective, *American Psychologist*, 47, pp. 723-729

Stevenson,, H. W., Chen, C. and Uttal, D. H. (1990) Beliefs and achievement: A study of Black, Anglo and Hispanic children, *Child Development*, 61, pp. 508-523

Stevenson, H. W. and Lee, S. (1990) A study of American, Chinese and Japanese Children, *Monographs of the Society for Research in Child Development*, No. 221, 55 (1-2)

Stevenson, H. W. and Stigler, J. W. (1992) *The Learning Gap: Why Our Schools Are Failing and What We Can Learn from Japanese and Chinese Education* (New York: Summit Books)

Stipek, D. and Gralinski, J. H. (1996) Children's beliefs about intelligence and school performance, *Journal of Educational Psychology*, 88(3), pp. 397-407

Tye, B. B. (1985) Multiple realities: A study of 13 American high-schools (Lanham, MD: University Press of America)

US Department of Education, National Center for Educational Statistics (1998) Pursuing Excellence: A Study of US Twelfth-Grade (Washington: National Centre for Education Statistics)

Wainer, H. (1993) Measurement problems, *Journal of Educational Measurement*, 30, pp. 1-21

Webber, S. K. (1996) Demand and supply: meeting the need for teachers in the 'new' Russian school, *Journal of Education for Teaching*, 22(1), pp. 9-26

Weiner, B. (1979) A theory of motivation for some classroom experiences, *Journal of Educational Psychology,* 71, pp. 3-25

Wilson, S. M., Henry, C. S. and Peterson, G. W. (1997) Life satisfaction among low-income rural youth from Appalachia, *Journal of Adolescence*, 20, pp. 443-459

APPENDIX TO CHAPTER 5

Illustrative examples of question types

Question type a

On the whole, how much do you like coming to school? *Please circle the appropriate answer*

- I like school very much
- I quite like school
- I don't feel strongly about school one way or the other
- I don't like school very much
- I dislike school a lot

Question type b

What are the most important reasons for success in school work *(please rank from 1 to 4; 1 being the most important, 4 the least).*

- Working hard
- Luck with test/exam questions
- Being clever
- Whether teachers like you

Question type c

Do you usually work as hard as you can?:

- In class   Yes/No *(please circle the appropriate response)*
- On your homework   Yes/No

Chapter 6

World Class Schools
Some Preliminary Methodological Findings from the International School Effectiveness Reserch Project (ISERP)

David Reynolds
Department of Education, University of Newcastle Upon Tyne, UK

Bert Creemers
GION, Groningen, The Netherlands

Sam Stringfield
Johns Hopkins University, USA

Charles Teddlie
Louisiana State University, USA

1. INTRODUCTION: THE RATIONALE FOR THE STUDY

Recent years have seen a greatly enhanced interest within educational research in issues of a 'comparative' nature. In most fields, there is an enhanced internationalisation evident in the increasingly international attendance at the conferences of educational 'specialties' and in the acceptance of the growing importance of 'nation' as a contextual variable.

The reasons for this internationalisation have been fully dealt with elsewhere (Reynolds and Farrell, 1996; Reynolds *et al.*, 1994) but briefly can be related to the ease with which ideas, both practical and conceptual, can now be spread throughout the world by information technology, and to

D. Shorrocks-Taylor and E.W. Jenkins (eds.), Learning from Others, 115–136.

the pressures that are being put upon national educational systems to maximise pupil outcomes by searching out and utilising effective practices from anywhere in the world that they may exist.

In this situation, the pressure upon the discipline of comparative education to resource this increasingly internationalised discourse has been growing, yet the discipline itself has been seemingly on an intellectual plateau for perhaps two decades. Much of the discipline appears to be simply descriptive, with little attempt to generate any theoretical underpinnings. There have additionally been 'macrolevel' attempts to generate theoretical understandings but these appear to be without empirical foundation. When there are discussions about educational policies in this literature, there is usually an assumption that the policies have the effects that their policymakers assume, a very dangerous assumption to hold! Perhaps most importantly, the conventional absence of any dependent variable internationally, across countries, against which to assess the various influences of the independent variables, makes analysis of the causal factors that determine the nature of countries' educational systems very difficult.

There has been, of course, an additional body of knowledge to resource the international discussion of education increasingly taking place, and that is the cross national achievement surveys undertaken by the IEA and similar organisations, such as the recent Third International Mathematics and Science Study (TIMSS) (see Chapter 2 in this volume). However, although they have the intellectual advantage of being able to compare countries on a common dependent variable or metric of achievement scores, the paradigm that the achievement surveys work within seems to be deficient in many important aspects.

1. These studies usually utilise cross-sectional samples of students, rather than following cohorts of students over time through longitudinal research designs.

2. The samples of students from countries are sometimes not representative of the entire populations of students in those countries, which is especially the case when specific geographic or political regions are chosen to represent an entire country.

3. Operational definitions of sampling 'levels' may vary across countries; for example, what constitutes third grade in one country may not be the same in another.

4. Materials may not be translated the same across countries, thus resulting in reliability and validity problems.

5. The cross-country reliability and validity of surveys and questionnaires may be suspect, since the constructs that they measure may be different in divergent cultural contexts; for example, Purves (1992) concluded that

written composition must be interpreted within a cultural context, not as a general cognitive ability.
6. There are difficulties in designing tests which adequately sample the curricula delivered in several countries.

Other criticisms of these international educational surveys address a different set of issues related to a perceived need for more 'contextually aware and sensitive' investigations of the processes at work in differentially effective schools. These criticisms include the following.

1. There are few 'data rich' descriptions of the school and classroom processes associated with more effective (and less effective) schooling in international studies (partly as a result of the use of questionnaires to gauge classroom processes rather than observations, no doubt because of cost considerations).
2. There is often a lack of information of non-educational factors (e.g. socio-economic status of students' families; community type in which the school is located) that may affect the methods whereby individual schools attempt to effectively educate their students.
3. There is often a lack of information on the affective or social outcomes of the schooling process.
4. There are few examples of international studies that successfully link 'country factors' (e.g. educational policies enacted at the national level, unique aspects of national educational systems) with the effectiveness of schools or with what factors appear to be effective within schools.

This, then, was the context in the fields of comparative and informational studies within which the International School Effectiveness Research Project (ISERP) was founded in 1991. It is important to note that not all the motivation for undertaking this project related to perceived deficiencies in the research base of comparative education paradigms, since the intellectual state of affairs in school effectiveness research was also a contributory factor.

Four sets of factors from the effectiveness community had an influence in leading to the setting up of ISERP. Firstly, the 1990s had for the first time seen research findings which suggested that the effective schools 'correlates' or 'factors' might be somewhat different in different geographical contexts, with the Dutch educational research community, particularly, being unable to replicate more than a handful of the 'classic' school or classroom factors from the American 'five factor' theories of educational effectiveness (see Scheerens and Bosker, 1997 for a summary). Particularly interesting and potentially important was the failure of Dutch empirical research to show the importance of the leadership of the Principal in creating effective schools (van de Grift, 1990), in marked contrast to the importance of this role and of the role occupants shown from American research (Levine and Lezotte, 1990).

The apparent 'context specificity' of school effectiveness factors suggested an interesting future direction for research, which would involve varying social context systematically between countries in order to see which factors universally 'travelled', and which factors did not but required particular cultural and social contexts to be potentiated in their effects.

Secondly, there was a clear need to generate theory, since the variation in 'what worked', if it could be explained and theoretically modelled, would force the field towards the development of more complex and multifaceted accounts than the 'one size fits all' mentality that had hitherto existed in the field. Useful contributions were already being made in the area of theory in the early 1990s (Creemers and Scheerens, 1989; Creemers, 1992) but there was a need to take them further.

Thirdly, it had increasingly become clear that the importance of comparative international research is that only these studies could tap the full range in school and classroom quality, and therefore in potential school and classroom effects. *Within* any country the range of school factors in terms of 'quantity' variables (such as size, financial resources, quality of buildings) and in terms of 'quality' factors (such as press for achievement) was likely to be much smaller than the variation in such factors *across* countries.

Fourthly, many effectiveness researchers had become concerned about the over-simple, not to say simplistic, potential transfer of educational policies that was already going on between countries. The 'importation' of educational policies from one country with one context to another country of different contextual conditions is of course something that has had a long history. However, when ISERP was being planned in the mid-1990s, the process had reached its peak in North America, with the popularity of the Stevenson (1992) cross cultural comparisons of Taiwan and the United States generating simplistic suggestions that the United States should adopt levels of children's time in school and the like that were characteristic of the Pacific Rim (see further elaboration in Reynolds and Farrell, 1996). Although in retrospect it is clear that the ISERP findings and speculations about Taiwan may well themselves have encouraged equally simplistic discussion in the United Kingdom (see Reynolds (1997) for a cautionary tale), it is sobering to remember that the possible explication of the *complexity* of these issues and the *difficulty* of cross cultural transfers was one of the principal reasons behind the intention to conduct ISERP.

ISERP eventually came to comprise nine countries, split across three broad regions: the United States and Canada; Australia, Hong Kong and Taiwan, and the Netherlands, Norway, the United Kingdom and Ireland. Each country selected either six or twelve primary (or elementary) schools, with these schools selected to be of high, average or low 'effectiveness', based upon either prior data on schools' intakes and outcomes or on nomination by those with close knowledge of the schools such as inspectors, advisers and the like. Half of the schools were selected from low socio-economic status communities and half from those of middle socio-economic status.

The basic methodology of the study was to follow a cohort of children aged seven through their schools for two years, the age being chosen to maximise the chance of ensuring that school effects did not masquerade as intake effects and because children in many societies of the world, such as Norway, did not begin their education until age seven. Mathematics was the dependent variable on which the relative progress of children and societies was to be measured, selected because it was more 'culture free' than any other measure given its similarity as a body of knowledge across cultures.

The ISERP study broke new ground in many areas. For student outcomes it utilised measures of children's social outcomes in the different countries, such as measures of attitudes to education, self-perceived ability in school and attitudes to teachers. Most innovative of all, the research included direct observation of teachers in different contexts in different countries, with the use of an adapted version of Robert Slavin's QAIT classroom observation system designed to measure quality of teaching, appropriateness of teaching, the teacher's use of incentives and the teacher's time use (Schaffer, 1994), to generate descriptions that were qualitatively rich on classrooms. The research design was also based upon the use of 'mixed' methods, both quantitative and qualitative. Immersion in the schools and cultures of different countries was to be undertaken by the 'core' research team, to understand the cultural 'taken for granted' that might be the explanation for country differences in the educational processes that appeared to 'work'. Rich case studies of schools were also to be undertaken.

There were two major research questions of the study.

1. Which factors are in general associated with student academic and social outcomes across countries and which factors are restricted to a certain cultural context?

2. Which factors are associated with student academic and social outcomes across countries, for students with different characteristics?

ISERP therefore involved trying to understand the effects of societal (country) differences on the ways whereby schools become effective and also investigated whether effective schools characteristics varied within and across countries depending on the characteristics of the students. Since the design involved case studies, it allowed for a better understanding of 'why' country differences in effective school characteristics might exist.

In summary, the characteristics of the ISERP research design in the early 1990s reflect what has by now become axiomatic within the school effectiveness research paradigm, in the following respects:

- the decision to focus on elementary grades, since it is easier to discern 'school effects' from 'intake effects' at the lower grade levels;
- the decision to look at context variables;
- the decision to use mathematics as the criterion variable;
- the decision to gather data at multiple levels – classroom, school, district – which has become axiomatic over the past decade as theory (e.g. Creemers, 1994; Scheerens, 1992) on educational effectiveness and mathematical models capable of analysing multiple levels of data have been developed (e.g. Aitkin and Longford, 1986; Goldstein, 1995);
- the decision to add affective and social outcomes;
- the decision to sample both outlier and average schools, which enables studies without huge resources to generate large sample sizes to 'maximise contrasts'.

2.   SOME FINDINGS AND LESSONS FROM ISERP

Analysis has only recently been completed and will be reported in full elsewhere, so this is not the place to give a full report upon our findings (preliminary statements are available in Reynolds and Farrell, 1996; Creemers and Reynolds, 1996). In general, the performance of countries within the ISERP group is very similar to results reported from other studies, with the Pacific Rim societies of Hong Kong and Taiwan having higher mean achievement levels. Interestingly, Anglo-Saxon societies such as the United Kingdom and much of the United States show a larger spread of scores, and a bigger standard deviation, than other societies, reflecting the contrast between the beliefs in Anglo-Saxon societies that everyone should be able to move at their own pace, with that in other societies that education should get all children 'over a hurdle' of basic skill achievement. Parental social class and educational level is strongly associated with children's achievements in Anglo-Saxon societies and more weakly associated in societies such as those of the Pacific Rim, reflecting probably the utilisation of a strong, common technology of schooling in the latter societies and the use of a technology that varies by social class background in the former.

Rather than pre-empt the final ISERP publications, what we would like to do in the remainder of this chapter is to outline some of the lessons from the ISERP study for the field of comparative education, based upon our experience of attempting to design and conduct a study with what we hope are the leading edge research principles outlined above. We begin with some methodological insights we think are of use.

2.1   Lesson One : Cohort studies are essential in comparative study

These studies are essential to understand the nature of the influences upon children's development in different countries since they permit the study of the student's change *over time* as well as their status *at a point in time*. Such cohort studies also control considerably more than cross sectional studies can for the influence of cultural, economic, social and other non-educational factors that may affect development, since these factors will affect children's initial levels on the tests utilised and be controlled out in their effects.

Cohort studies also permit a dynamic analysis of how children interact with their educational settings over time, furthering the understanding of how development is undertaken.

In the case of the ISERP study, it was the cohort design that helped us discover the various process factors of different systems that were in evidence in such areas as:

– the transitions between years within schools, which were managed much better within some countries than others;
– the ways in which children who presented problems educationally in some societies were allowed to 'cascade', whilst in others these were truncated before they were allowed to develop;
– the ways in which teachers developed over time in different countries.

More generally, it is clear that the way to understand how children develop over time in different country contexts is to study how the children interact with the variable influences, which cannot be shown by cross-sectional, at a point in time, studies.

2.2   Lesson Two : Gathering observational data on classrooms is essential – but difficult

In the last five or six years it has become close to axiomatic within school effectiveness research that the classroom or 'learning level' explains far more variance in children's achievement than the school level (Creemers, 1994; Teddlie and Reynolds, 1999). However, historically international studies have focused heavily upon the school level for the collection of educational process data, rather than classrooms.

It is clear from our own analyses that one can only understand the dynamic interaction between children and educational systems by understanding education as experienced by children at the point of delivery. Utilising measures of teachers' instructional behaviours in lessons, as well as the technique of following a typical child through a whole school day as experienced in classroom, proved to be amongst the most powerful of our methodological techniques, since it focused attention upon instruction of children and gave a well-grounded sense of life in schools.

There are, though, notable difficulties in conducting classroom observations in different countries. Training of classroom observers in different countries is clearly essential, given the likely wide geographical and cultural spread of the countries and the difficulties of controlling any range of countries in their data collection once started. Training and estimation of reliability is also essential if any use is to be made of the observational data to explain country differences in achievement scores, since differences between countries need to be securely related to educational reality rather than to differences in the judgements of observers. However, training in classroom observation is clearly expensive and time consuming, and even the ISERP study did not undertake enough of it.

Even considerable training would not have got around the problem of differences between countries in what the core constructs of the observation instruments actually 'meant'. The observation instrument was partially one of *high* inference measures, and partially one of *low* inference. The measurement of pupils' time on task was an extremely low inference measure in which observers simply looked around a class every eight minutes or so, and judged what proportion of the class were 'on task', defined as working, listening or in other ways concentrating upon their class task. However, the measurement of such aspects of the classroom as the quality of the teachers' behaviour proved more difficult, since it required the gradation of teachers in terms of aspects such as their 'classroom control' techniques, their 'exhibiting of personal enthusiasm', or their 'skill in utilisation of questioning', all areas in which their behaviour was clearly being judged by 'high inference'. In Taiwan, for example, it is clear that the concepts of 'positive academic feedback' and 'negative academic feedback' caused particular problems given Taiwanese definitions of 'positive' and 'negative'. This scale item had been chosen with an Anglo-Saxon or European conception in mind, in which positive feedback would be a statement by the teacher such as 'well done – keep it up!' and negative feedback would be evidenced by a teacher saying something such as 'that's poor work – you need to improve'. In Taiwan, and to an extent in Hong Kong, such judgements proved more difficult, since it was possible within the culture to see extreme negativity from a teacher as being 'positive', and likewise an attempt to shield the child from criticism by saying something pleasant about their work as 'negative academic feedback', given the overarching national need to achieve high levels of academic performance.

Although we clearly encountered difficulties in our use of observational methods of classroom processes, their use is considerably superior to the continued use of questionnaire methods of data collection on classroom and school processes that is still part of what we noted above as the 'IEA tradition'. In the recent TIMSS study of variation in mathematics achievement, for example, (Keys, Harris and Fernandes, 1996) the only possible explanation for the very low proportion of children from Pacific Rim societies reporting that they were 'very rarely tested' was that the possibility of a 'test' being seen as separate from the conventional routine of instruction was not something that was in the mind set of children filling in the questionnaires. Even the most speedily trained observers were able in our study to distinguish 'testing' from 'teaching' in all our countries, giving us probably much more valid data to use to generate useful explanations than those derived from questionnaires.

One useful way forward for classroom observation may be the use of the 'whole school day methodology' (WSD) or 'child study' (Schaffer, 1994). In this process, one 'typical' child from the study cohort at a school was selected, based on nomination from the school's staff. That child was then followed by a researcher for an entire school day, a process also known as 'shadowing'. The process entails sitting as unobtrusively as possible in the classroom during the school day, following the target child to other classes and experiences. Notes taken were open-ended and anecdotal, focusing on the child's actual activities, participation and level of engagement in class, and the extent and type of interactions with other children and adults.

The strength of the WSD observations lie in the extent to which they bring the children's experiences of schooling to life. Our researchers responded to two general prompts when conducting the WSD.

– What are the events in the life of a typical student at the school on a typical day?

– Does the school experience seem to make sense to the child?
  According to Schaffer (1994) the functions of the WSD are:

– to focus classroom observers' attentions on the impacts of instruction on individual students;

– to provide any naïve observer with a much richer, more grounded sense of 'life in the schools';

– to describe aspects of school experience that affect individual children, including patterns of instruction, access to knowledge and instructional quality;

– to examine instruction as received, that is from the child's perspective.

Many of our researchers reported that the WSD was one of the most illuminating of all the data collection procedures utilised in the case studies. This appeared to be especially the case for novice researchers, who had not had recent experiences observing in schools. Focusing on one student both simplified and intensified the observational experience for these researchers.

2.3   Lesson Three : Utilising social outcome measures is essential – but difficult

We had originally decided to broaden the range of measures from academic achievement only that we utilised as the outcome of educational processes at the end of year one and two of the study, to include some social or affective outcomes such as children's 'locus of control' or attitudes to education. The intention was literally to force ourselves to generate more rigorous 'theories' because of the likelihood that different countries would generate different patterns of relationships for the different outcome data, with some countries performing better on social outcome data than they did on academic outcomes.

We had only limited success in realising the promise offered by multiple outcomes, however. Firstly, the amount of variance related to educational factors on the attitudinal measures was limited (3-4 per cent), and less than that explained on the achievement measures. Secondly, it became clear that countries differed far more on what were their national educational definitions of appropriate social goals for schools, than on their academic goals. As an example, the Norwegian educational system evidenced a unique commitment to 'democratic values' of involvement, activity in the community and participation in the life of the state, which clearly generated a need to tap these for all students in all countries in order to be able to study them. However, these concepts or constructs of 'democratic values' had no apparent existence within the educational systems of any of the other eight societies and particularly not in the two systems of the Pacific Rim. Any decision, therefore, to measure non-academic outcomes requires a substantial investment of effort to establish exactly what the non-academic goals of the various countries being studied are, before the equally time-consuming task of instrument development.

There are, though, numerous difficulties then involved in the construction and utilisation of measures in the affective or social areas. Outcome measures in areas such as mathematics are not dependent upon the perceptions that exist in different countries - two plus two equals four in all the nine countries that we studied, as it were. However, some social concepts did not 'travel' in the same way across various countries, such as the locus of control measures derived from the American literature (Crandall *et al.*, 1965). Anglo-Saxon thinking on this issue posited the existence of children who could be scaled as either internally controlled (seeing themselves as responsible for their own success or failure) or externally controlled (seeing themselves as determined by broader forces that were largely out of their control). Children were accordingly given situations in questionnaires in which they might find themselves, and asked to choose either an 'external' or 'internal' response as the reason for the existence of the particular situation. Such attempts to use oppositional, 'either/or' categories simply generated confusion and mystification amongst the children of the Pacific Rim, whose Confucian cultural traditions included both a stress on the importance of an individual's effort and striving *and* an emphasis upon the influence of religious and mystic forces outside the control of the individual, both operating at the same time!

2.4   Lesson Four : Mixed methods are essential in comparative work

We had originally decided to utilise multiple methods of investigation rather than *only* quantitative or qualitative methods, in a pragmatic decision to be guided by the needs of answering our research questions rather than to reflect any educational philosophies. Quantitative data was clearly essential in order to understand the country and school performances on the 'dependent variable' of mathematics achievement, yet qualitative case study data obtained from participation in the educational life of different schools in different countries was essential in order to understand the educational processes.

After conducting the research, it is clear that the use *only* of quantitative *or* qualitative data would have been highly restrictive in terms of the kinds of understandings that we could have developed. Only quantitative data would have given us the dependent variable of pupil performance, and those school and classroom variables that could have been expressed in numerical data, such as teacher behaviours, school size and indeed any factors that could be measured and quantified through questionnaires and structured interviews.

However, it was the qualitative data collected through 'rich immersion' in school and classroom life that shed light upon many of the differences between countries that we were interested in. As an example, we utilised a programme of inter-visitations in which observers from the 'core' organised a process of 'cultural immersion'. Such an inter-visitation had the following purposes:
1. to triangulate or verify observations made by the local country teams; and
2. to provide a more distanced, often quite different, lens through which to view the local/national scene.

The ideal inter-visitation process in ISERP involved the following steps:
1. visitors from one country (ideally a two person team) requested to be allowed to visit representative schools in the ISERP sample in another country;
2. the host country agreed to the inter-visitation and established a schedule of visits to schools (ideally three to four schools in different SES and effectiveness categories) in their country;
3. the visitors spent at least a day in each school, making informal observations and interviewing faculty and students at the schools, using the host team as translators if necessary;
4. the visitors described their initial impressions to the hosts before exiting the field, and the hosts reacted to these impressions;
5. the visitors wrote a report summarising their impressions, including differences that they perceived between the schooling processes in the host country and their own country;
6. the host country team responded to the inter-visitation report, noting areas of agreement and disagreement; and
7. the inter-visitation documents became part of the qualitative data base used to generate our case studies and conclusions.

As an example, the American visiting team produced 30 page reports from both its Dutch and Norwegian visits. These reports had the following outline.
1. General Differences Between Schools in the US and the Netherlands (Norway).
2. Descriptions of Schools Visited in the Netherlands (Norway).
   e.g. School A
      1. School Context
      2. School Level Impressions
      3. Classroom Level Impressions
      4. Student Level Impressions

The following are excerpts taken from the first section of the reports (General Differences Between Schools in the US and the Netherlands or Norway).

On several potentially important dimensions, variance among the Dutch schools was less than we would have expected to find among US schools. There appeared to be more of a collective approach to schooling in the Netherlands than is found in the United States. This may be partially a function of the particular schools we visited. Nevertheless, we perceived that on such traditionally important dimensions as school goals, there was a shared clarity of vision among the Dutch schools visited. For instance, all children were supposed to begin reading *together* at Level 4 (age 7).

Head-masters (principals) in Norwegian schools seemed to have less power in determining the goals and general direction of their schools than principals have in the US (i.e. if the US principals choose to use it). Perhaps it is simply that they choose to share power with their teachers and community more. Part of it would appear to be a conscious choice on the part of government to set up individual school steering committees on which the principal has no formal vote.

2.5 Lesson Five : The educational factors studied in comparative research must reflect practice in all societies – but this is difficult if not all societies have contributed to existing research and conceptions

The ISERP enterprise that went into the field to study school and classroom processes was ethnocentric in its views as to which factors should be studied. In our favour, we *were* gathering data on the multiple layers of the educational system, in the areas of Districts/LEAs, schools and classrooms. However, the conceptualisation of which school effectiveness and teacher effectiveness factors were to be measured within the nine different countries reflected a literature review about 'what matters' in affecting educational achievement that had been conducted by the 'core' team (Reynolds *et al.*, 1994b). However, this review of course mostly reflected the concerns of the countries that had contributed to the school effectiveness literature, namely those Anglo-Saxon societies within the United States, the United Kingdom and Australia. Little attempt was made to include school effectiveness factors that might have been important within the *different* cultural contexts of other societies in which there had *not* been school effectiveness research. Vulliamy (1987), for example, noted the importance of contextually specific factors in Papua New Guinea in accounting for whether a school was effective or not, in this case 'an absence of corruption in teaching appointments to the school' which turned out to be heavily associated with effectiveness.

Additionally, many of the factors customarily used within the research literatures of 'Anglo-Saxon' societies seemed to have very limited explanatory power within societies like the Pacific Rim for example. The personal and biographical variables that often explain variance in effectiveness studies within the Anglo-Saxon societies reflect the importance of personal factors in determining the quality of individual teachers' and head-teachers' practice, because practice is personally determined. One would expect such factors to be much less important in Pacific Rim societies where the giving of a 'strong technology' of agreed practice to all teachers and head-teachers is explicitly designed to eradicate such personal influences.

Further examples of the need for cross-national relevance in the resources utilised are provided by the 'intake' or 'control' variables used in our study. Occupational classifications derived from the status hierarchies present within Anglo-Saxon societies may explain less variance when utilised within Pacific Rim societies that have different definitions of which occupations have 'high' and which 'low' status, and where different kinds of stratification (e.g. between urban and rural populations) are salient.

2.6   Lesson Six : There are few agreed international constructs concerning effectiveness

When we began to design ISERP, we had, as a group of Anglo-Saxon scholars, a clear intention to expand the number of societies that were to be studied in order to expand the variance on the dependent variables (such as mathematics achievement) from the constrained variance that was in evidence within Anglo-Saxon societies, and on the independent variables (such as school and classroom processes) likewise.

However, it came as a considerable surprise to discover that many of the basic constructs of the Anglo-Saxon school effectiveness movement were only in existence in roughly half of the countries under study. In Taiwan and Hong Kong there was no discourse about 'school effectiveness' or 'teacher effectiveness', separate from or distinctive from the general societal commitment to ensuring that all children learn. Neither in these societies, nor in Norway, was there any understanding of the effectiveness notion of school 'value added', and effectiveness of institutions was seen as related only to the 'raw scores' or 'raw achievements' of children, not to the relationship between their intake and their outcome scores.

Table 6.1 shows the dimensions that were produced by the 'core' team to provide a framework for the collection of data from the country case studies of their schools, as at the beginning of the study. The indicators for the various dimensions clearly relate much more closely to the discourse about school effectiveness present in countries such as the United Kingdom and the United States than in others. There are, and this needs reflecting in research designs, few agreed constructs concerning effectiveness that are common internationally, suggesting that, in future international studies, it may be best to have both a common core of process factors that are seen as potentially useful across all country contexts (e.g. expectations of academic achievement) and additionally have 'country specific' factors that are only utilised in the search for useful explanations in certain countries (e.g.

'image' in Anglo-Saxon countries, the students relations with others in Taiwan and/or the guidance/counselling system in Norway).

Table 6.1. Twelve Case Study Dimensions with Prompts, 1992

Case Study Dimension	Prompts
(1) General Characteristics of the School	Student Body Size Number of Teachers and Staff Description of School and Catchment Area Public/Private Organisation
(2) The Child's Experience of Instruction (The Child Study)	Events in the Life of a Typical Student at the School on a Typical Day Does the School Experience Seem to Make Sense to the Child?
(3) Instructional Style of Teachers	Teacher behaviours in classrooms
(4) Curriculum	What is Taught? When is it Taught? Techniques for Teaching Math, Etc.
(5) Influence of the Parents	Contacts with the School How the Parents Participate, Etc.
(6) Principal	Style of Leadership Organisation of the School, Etc.
(7) Expectations	Expectations for Students Assessment System Discipline System, Etc.
(8) School Goals for Students	Academic Goals Social Goals, Etc.
(9) Inter-Staff Relations	Contact Between Staff Members How is Information Passed on about Students? Etc.
(10) Resources	The Source for the Resources How Much Outside Resources Are There? How Important Are the Outside Resources? Etc.
(11) Relationship with Local Authorities/ Districts	What Type of Contacts? How Important Are These Contacts? Etc.
(12) School Image	School Presentation to the Community Is an Image Necessary for the Survival of the School? School Atmosphere, Etc.

Table 6.2 shows the responses by representatives of the different research teams towards the end of the ISERP study to a request to elucidate the key, or most important, factors that were usually held in their countries to be the characteristics of various classic school effectiveness dimensions. Certain educational characteristics, such as the 'image' of schools in affecting their effectiveness, cannot be described by certain country teams (such as Norway) since they have no meaning or relevance in the culture. Where

certain educational features *are* a part of educational discourse, such as in the areas of 'Teachers Instructional Style' or 'School Goals', it is notable that there are considerable differences in what each country regards as the salient and important features in the particular area that might be influencing effectiveness. In the case of 'Teacher's Instructional Style', note how in Ireland the key factors in this area are whether teachers are 'progressive' or 'traditional' in methods and whether they are 'organised' or 'disorganised' in their practice, whereas in Taiwan the most important issues concern whether they are 'innovative' in their methods and 'with it' in their interaction with pupils.

Table 6.2. Detailed Description of Effectiveness Constructs by Different Countries, 1995

	CANADA	USA	NORWAY	UK	IRELAND	TAIWAN	N'/LANDS
School Characteristics	- size - location	- SES - characteristics of students - physical .condition of schools	- catchment area/demographic/ethnic stability	- size - resources per head - staff turn-over and characteristics	- physical building/ location - pupil intake - teaching staff characteristics	- physical upkeep - neighborhood - history	- characteristics of student population (IQ) - denomination (Cath. public etc.)
Child Experience	- self concept - teacher-child relationship - teacher personality	- degree of fragmentation of day - is school a coherent experience - teacher links child's experience to academic lookout - level of academic focus of day	- motivation - learning strategies	- school climate - lesson transitions consistency of learning experience - coordintion of instruction	- time factors - routine - pupil involvement in decision making - social relationships	- student behaviour - interaction,	- degree of structure of school day - motivation - speaking Dutch with parents - high involvement in activities - positive teacher - feedback
Instructional Style	- collaborative - problem-based - teacher-directed	- classroom climate - management of class - skills of inst. delivery - use of innovative practise	- Interactive teaching - group - whole class - soc/academic focus - guidance	- proportion of whole class direct instruction - teacher man. of groups - time use - interactive teaching - match of task/pupil	- traditional progressive - organised/ disorganised	- inventive methods - curriculum match - interaction with-it-ness	- no of disc. problems - whole class teaching - time use - feedback - assessment - whole class instruction - high ins. tim - eff. class management
Curiculum	- locally developed - outside mandated	- fac input - degree of adoption of the state initiatives at school level	- coherence between national and local - textbooks	- use of homework - quality of instructional material	- centrally dictated - school plan - extra curricular activities	- adherence to Central Plan - special characteristic of school	- workload per day - curr. fits with students - curr. content - use of homework - remedial material - content per grade - high OTL

	CANADA	USA	NORWAY	UK	IRELAND	TAIWAN	N'/LANDS
Parental Influence	- role in school	- helping child with homework - attending PTO meetings - raising money for school - choosing a neighbourhood that has good school	- formal/informal - degree of decision-making	- participation in activities - formal organisation of parents - informal treatment of parents	- school level/ fundraising/ parent ass. - class level/ inv. in teaching act. - home/school liaison	- cultural limitations - financial support	- parental participation - agreement with parents
Principal	- educational leadership - personal leadership - administrative leadership	- style top/down, bottom/up - man daily routine comm. to instructional process - selection/ replacement of teachers - guidance of school into new programms	- internal org. and decision-making - planning and flexible scheduling leadership style - administration daily life - comm. to instruction	- involv. of staff - commitment to instruction - proactive staffing - active staff development	- style of leadership - admin. function - educational leadership	- knowledge - awareness - thoughtful-ness - attempt at improvement	- N/A
Expectations	- teacher exp. for individual accomplish-ments - teacher exp. for behaviour	- exp. for academic achievement - exp. for behaviour in the school	- change to assessment - form/inform. assessment - student assessment - tests - reward syst. - teacher appraisal	- teacher exp. of each other - exp. of academic achievement - exp. of behaviour	- personal/ social development - academic learning	- academic - life skills - relations with others	- academic press - clear consequences of negative behaviour
School Goals	- existence of formal school goal setting - use of formal goals in daily interaction	- focus in adademic goals - focus on increased parental mv. - focus on staff development - focus on total development of the child	- written goals in school plan - conn. with priority areas for innovation - formal and informal goals	- number - focus on academic achievement - focus on social goals	- explicit - implicit	- for students - for all schools - expanded goals at other schools	- focus on academics - focus on climate - focus on academic achievement - clear final aims - homework policy
Inter-Staff Relations	- team curr. development - team teaching - collegiality	- collegiality/ among staff - collective academic planning - phys. plant - techn. in classrooms - extra financ. resources for enhancement - availability of material	- professional relations - social relation	- professional support - personal support	- professional co-operation - levels of friendship - quantity/ availability - maintenance - upkeep	- physical structure of office space - instructional leader role	- number of part-time teachers - cohesion with respect to instruction
Resources	- pupil-teacher ratio - access to facilities - access to materials	- phys. plant - technology in classrooms - extra financ. resources for enhancement - availability of materials		- building and physical plant	- quantity/ availability - maintenance upkeep	- contribution or use	- N/A

	CANADA	USA	NORWAY	UK	IRELAND	TAIWAN	N'/LANDS
Assessment	- evaluation	- tests of student performance - teacher evaluation - program evaluation	- standardized distr/state assessment - school generated assessment practices	- standardised evaluation policy - classroom evaluation policy - linking student evaluation with managerial decisions	- N/A for pupil testing - individual record keeping	- by local authorities	- school evaluation policy - standardised testing evaluation policy
Image	- role of image in parent choice of school	- image in parental community - image in academic community	- N/A	- management - image in the community - use of visitors	- image in local community	- community - with parents - with students	- N/A

2.7   Lesson Seven : In spite of likely difficulties, comparative education research can generate large increments in understanding

The ISERP study has validated for those of us involved in it the value of comparative study, for three reasons. Firstly, we have simply seen in other societies a variety of educational practices at classroom and school levels that would not have been seen had the core research team stayed within their own societies. In Pacific Rim societies for example, the majority of lesson time is filled with what has been called 'whole class interactive' instruction, in which relatively short lessons of 40 minutes are filled with fast, emotionally intense presentations from teachers, with accompanying very high levels of involvement from pupils. This model of teaching, which is also found within a European context in societies such as Switzerland, is now of course the subject of considerable debate within a number of countries e.g. the United Kingdom. In Norway, as a contrast, there is no formal assessment of children through the entire phase of their elementary/primary education from the age of seven, a marked contrast to the United Kingdom practice of formal assessment and associated publication of results. In Pacific Rim societies again, one can see micro-level educational practices such as teachers teaching from a stage at the front of the class some six inches high (to help those at the back of the class to see), pupils marching to assembly through corridors in which loudspeakers play music (to ensure a relaxed attitude) and pupils starting the afternoon session of school with a 'sleeping' lesson (to deal with the fatigue brought about by the frantic pace of the school and the heat/humidity of the climate). Put simply, comparative investigation shows an enhanced range of what appears to be educationally possible.

The benefits from comparative investigation are more than simply a knowledge of educational factors that might of course be utilised in programmes of experimentation in one's own country. They are, secondly, that one is made aware of educational philosophies that are radically different from one's own, or those of the government of one's own country. In Norway, for example, there is a strong commitment to the child as an 'active citizen', and to what are called 'democratic values' that have no British or American equivalents. In Pacific Rim societies like Taiwan, there is a philosophy that the role of the school is to ensure that all children learn, and that a strong 'technology' of practice should be employed to ensure that children are not dependent on their family background. Such societies are very concerned about the use of practices to improve the achievement of their trailing edge of pupils, therefore, and are rather less concerned with the education of the 'gifted and talented' than are the societies of the United Kingdom and United States.

There is a third reason for comparative investigation that is probably even more important than the two above, concerning the possibility that within the right kind of comparative framework one can move beyond looking at the practices of other societies and actually so empathise with other societies that one can look back at one's own society with the benefit of their perspective. Such 'acculturation' is what happened to many of us in ISERP when we were confronted with, and may have identified with, Pacific Rim educational systems. Looking back at the British system through their 'lens', one wonders at the utility of the combination of the very complex technology of practice that is evident in British primary practice, for example, with methods of teacher education that are premised on the importance of teachers 'discovering', or at the least playing an active role in learning about, the appropriate methods to use. To a Taiwanese educationist, this celebrates the desires of teachers for their long-term developmental needs above the needs of children to receive a reliable, consistent, predictable and competently provided experience as they pass through their schools.

The use of another culture's 'lens' adopted through the inter-visitation programme to better understand the limitations and strengths of one's own educational practice applies at the level of educational philosophy as well as educational practice. As an example, those of us involved in the British ISERP team would have historically viewed our primary education practice as loosely 'progressive' and indeed would have thought that in many senses it was the envy of the world. The encouragement of children to learn on their own rather than simply being instructed, the new sets of social outcomes that the system is widely argued to concentrate upon and the reduced emphasis upon the testing of knowledge acquisition have been widely argued to be the hallmarks of progressive practice in the British system.

Seen from a Pacific Rim perspective, however, the characteristics of the British system would be seen as regressive, not progressive. Transferring the burden of learning to pupils would be seen as maximising both social class influences and variation between pupils within Taiwanese educational culture, since pupils' learning gains would depend on what they brought to the learning situation in terms of achievement levels and backgrounds. Removing the 'constant' of the teacher would be seen as further maximising individual variation in knowledge gain. Avoiding the testing of basic skills could be seen as maximising the chances of children who have missed acquiring particular knowledge bases being left without them, through the absence of short term feedback loops that alert school authorities that certain children have not learned.

3.   CONCLUSIONS

We do not have space in this chapter to discuss in detail the substantive intellectual lessons we have learned from our data, although they are probably even more powerful than our methodological lessons. Substantively, it seems that:

– large advantages accrue to those cultures who can generate 'strong systems' of education, as against those societies that rely on the necessarily finite numbers of people with the personal characteristics to generate effective practice;

– large advantages accrue to those societies where there is cohesion amongst education 'levels' in their attitudes to educational matters.

We also possess much fascinating material on the factors at school and classroom level that are associated with effectiveness within and across countries, which will soon be in the public domain.

However, we can conclude that comparative study, particularly if it reflects the lessons that we ourselves have learnt through conducting the ISERP study, has huge potential to advance our understanding in spite of the very clear difficulties involved methodologically. Comparative research expands variance on dependent and independent variables. It permits societies to see practices that are unknown within their own educational systems. Indeed, our own experience is that it can advance thinking about educational matters in ways most unlikely in 'within nation' research.

Given these huge benefits, it remains a great pity that comparative education in general, and the international achievement surveys in particular, have not shown consistent improvement in the quality of the data gathered, and in the insights derived from that data, over the last 20 years. In the absence of an intellectually vibrant comparative education community, the increasing tendency of educational research to be cross-national or international in focus will not be resourced, and the sub-disciplines of education may make the kind of intellectual and practical errors that comparative education could have warned them about.

Within comparative education, it is perhaps the large-scale cross-national achievement surveys that have most to offer, were they to improve in quality. These surveys are well known by educational researchers and policymakers, and command attention because of the themes they address. For all their faults, they have a common dependent variable and therefore, in theory, can handle the explanation of the effects of different patterns of independent variables. They include material on the focal concerns of educational research – schools and, to a more limited extent, classrooms. From our own experience, the quality of this work would improve if it, to summarise:

- utilised multiple methodologies and strategies of data collection;
- focused upon classrooms more, utilising observation of children's whole school days;
- adopted multiple outcomes;
- was sensitive to the variation between countries in their basic educational discourses;
- ensured that the factors studied were representative of the likely causal factors across all countries;
- utilised cohort studies that kept researchers in touch with the same children over time;
- utilised inter-visitations across countries to understand educational phenomena better.

REFERENCES

Aitkin, M. and Longford, N. (1986) Statistical modeling issues in school effectiveness studies. *Journal of the Royal Statistical Society, Series A*, 149 (1), pp. 1-43

Crandall, V. C., Katkovsky, W. and Crandall, V. J. (1965) Children's beliefs in their own control of reinforcements in intellectual-academic situations. *Child Development*, 36, pp. 91-109.

Creemers, B. P. M. and Scheerens, J. (eds) (1989) Developments in school effectiveness research. A special issue of *International Journal of Educational Research*, 13 (7), pp. 685-825.

Creemers, B. (1992) School effectiveness and effective instruction - the need for a further relationship. In Bashi, J. and Sass, Z. (eds), *School Effectiveness and Improvement*, (Jerusalem: Hebrew University Press)

Creemers, B. (1994) *The effective classroom* (London: Cassell)

Creemers, B. P. M. and Reynolds, D. (1996) Issues and Implications of International Effectiveness Research. In *International Journal of Educational Research*, 25 (3), pp. 257-266.

Goldstein, H. (1995) *Multilevel models in educational and social research : A Revised Edition* (London: Edward Arnold)

Keys, W., Harris, S. and Fernandes, C. (1996) *Third International Mathematics and Science Study, First National Report, Part 1.* (Slough: NFER)

Levine, D. U. and Lezotte, L. W. (1990) *Unusually effective schools : A review and analysis of research and practice* (Madison, WI: The National Center for Effective Schools Research and Development)

Purves, A. C. (1992) *The IEA Study of Written Composition II : Education and Performance in Fourteen Countries* (Oxford: Pergamon Press)

Reynolds, D., Creemers, B. P. M., Stringfield, S., Teddlie, C., Schaffer, E. and Nesselrodt, P. (1994) *Advances in School Effectiveness Research and Practice* (Oxford: Pergamon Press)

Reynolds, D. and Farrell, S. (1996) *Worlds Apart? – A Review of International Studies of Educational Achievement Involving England* (London: HMSO for OFSTED)

Reynolds, D. (1997) Good ideas can wither in another culture. *Times Education Supplement*, 19th September, p. 2

Schaffer, E. (1994) The contributions of classroom observation to school effectiveness research, in Reynolds, D., Creemers, B. P. M, Nesselrodt, P., Schaffer, E., Stringfield, S. and Teddlie, C. *Advances in School Effectiveness Research and Practice.* (Oxford: Pergamon)

Scheerens, J. (1992) *Effective schooling : Research theory and practice.* London : Cassell.

Scheerens, J. and Bosker, R. (1997) *The Foundations of School Effectiveness.* Oxford : Pergamon Press.

Stevenson, H. W. (1992) Learning from Asian Schools. *Scientific American,* December, pp. 32-38

Teddlie, C. and Reynolds, D. (1999) *The International Handbook of School Effectiveness Research* (Falmer Press: London)

van de Grift, W. (1990) Educational leadership and academic achievement in secondary education. *School Effectiveness and School Improvement*, 1 (1), pp. 26-40

Vulliamy, G. (1987) School effectiveness research in Papua New Guinea, in *Comparative Education*, 23 (2), pp. 209-223

Chapter 7

Making Use of International Comparisons of Student Achievement in Science and Mathematics

Edgar Jenkins
Centre for Studies in Science and Mathematics Education, University of Leeds, UK

Comparative studies in education, both within, and between, educational systems are by no means new, although the nature and focus of the work undertaken have changed significantly over time (Noah and Eckstein 1969). For much of the second half of the twentieth century, the emphasis has been upon understanding the relationships between educational practice and the social, political and historical forces at work in the wider society. This work, essentially qualitative and hermeneutic in nature, is typified by the work of scholars such as Hans (1950) and Kandel (1955). A different tradition, which, arguably, has now come to dominate the field of comparative studies of mathematics and science education, makes much greater use of quantitative data to correlate 'outcomes', notably student achievement, with a range of 'inputs' such as student characteristics or teacher quality, or with mediating 'processes' such as the form, content and structure of, and the emphases within, the mathematics or science curriculum and the teaching methods deployed. At the end of the twentieth century, qualitative and quantitative approaches to comparative research both remained in evidence, although neither had escaped unproductive controversies about the relative merits of each which has marked the wider social science literature. For some, an empirical approach is a form of positivism that ultimately attempts to quantify the unquantifiable. For others, a failure to adopt such an approach can lead only to, at best, unwarranted correlation and, at worst, unfounded speculation. Unsurprisingly, the best comparative work is governed by an acknowledgement that the research method or methods to be used are those which are most likely to lead to answers to the questions that have been posed.

D. Shorrocks-Taylor and E.W. Jenkins (eds.), Learning from Others, 137–157.

Even with this acknowledgement, however, formidable difficulties remain in undertaking comparative studies within education. Some of these difficulties, such as the high cost of international studies and the different ages at which pupils start formal schooling, can be overcome, or at the least be minimised, by adequate resourcing and research design. Others, such as ethnic, linguistic or other types of bias in test instruments or agreement about the variables between which correlations are to be sought, are both more subtle and more problematic. These methodological and other difficulties are important not just for the conduct of comparative research. They have a direct bearing upon the issues with which this chapter is concerned, namely the use that can properly be made of comparative data in science and mathematics education. Comparisons of student achievement based upon data from tests that show a strong cultural, linguistic or other bias are likely to be invalid. Likewise, comparative accounts of science or mathematics curricula or of the methods used by teachers to teach a given topic will be of limited use if they do not take account of the wider educational, social and other factors that influence curriculum construction in different contexts and help to shape teachers' everyday working practices and, beyond this, the attitudes of students and their parents to schooling.

Given these concerns, it is not surprising that international comparisons of student achievement in science or mathematics, along with the use made of them, have been the focus of considerable criticism. This has been particularly the case for some of the earlier international comparisons which were unable to accommodate adequately the curriculum complexity found among the different educational systems being studied and/or to distinguish satisfactorily between the intended curriculum and the curriculum as 'delivered' by teachers in the schools. Ultimately, however, there is an insuperable difficulty here. This is that the construction of a common set of test items which is then deployed to assess the performance of students in different educational systems or contexts serves to define a *de facto*, notional standardised curriculum.[7.1] The test results take no account of the relationship of this notional 'international' curriculum to the curriculum associated with each of the individual systems or contexts involved in the comparison. It thus ignores issues such as the history of individual science or mathematics curricula and the professional, social and pedagogical characteristics associated with the teaching of those curricula. To express this point in a rather different way, no international comparison of student achievement sheds much light on how well students within a given educational system understand the science or mathematics as taught to them

---

[7.1] From this perspective, the growth of international comparisons of student achievement in science and mathematics can be regarded as contributing to the globalisation of science and mathematics curricula.

within their own educational system. An important corollary is that there are severe difficulties in trying to establish what constitutes 'good practice', i.e. practice that is effective in promoting student learning, in an international context. The notion of *best* practice, as applied to pedagogy, is, of course, a chimera that serves, among much else, to close debate and render unnecessary any further research or inquiry into either the teaching or the learning of science and mathematics.

The Third International Mathematics and Science Study (TIMSS), (see Chapter 3), attempted to address some of the concerns expressed above by augmenting the international tests of student performance with questionnaires, a videotape classroom study (eighth grade mathematics only), a survey of science and mathematics opportunities (six countries) and an analysis of curriculum intentions (the Curriculum Analysis study). TIMSS involved approximately half a million students from 41 countries, with tests making up a total of 30 different languages. The outcome is an immense volume of qualitative and quantitative data, capable of being analysed in a number of different ways and of sustaining rather different conclusions. Relevant earlier studies include the First (FIMS 1964) and Second Mathematics Studies (SIMS 1982-3), the First (FISS 1970-72) and Second Science Studies (SISS 1984) sponsored by the IEA, the IEAP mathematics and science studies in 1988 (IAEPM1 and IEAPS1) and 1990 (IEAS1 and IEAPS2), and a number of other smaller scale investigations involving, typically, two, three or four educational systems. The early years of the twenty-first century will also see the first results of the Programme for International Student Assessment (PISA), carried out under the auspices of the Organisation for Economic Co-operation and Development (See Schleicher's contribution to this volume).

1.   ARE INTERNATIONAL COMPARISONS OF ANY USE?

Given the difficulty, complexity and cost of undertaking international comparisons in science and mathematics education, the volume of scholarly criticism that has been levelled at much of the work that has been done and the substantial reservations needed in interpreting the outcomes, it is important to ask whether such outcomes are of any use. The answer would seem to be yes, subject to two qualifications.

The first is that the limitations of such comparisons be fully recognised so that hasty inferences and naïve, if beguiling, conclusions can be avoided. The limitations are of various kinds and many are described, and their significance frequently contested, in the research literature. Some of the limitations relate to the aims of the comparisons undertaken. Orpwood, in this volume, identifies three different statements of purpose for the Third International Mathematics and Science Study (TIMSS) and argues that this diversity has had a significant influence on how 'several key aspects of the project were carried out' and thus on the kinds of outcomes that were obtained and the uses that can legitimately be made of them. From the point of view of the 'end-user' of data generated by international comparisons of student achievement in science and mathematics, Hussein poses the question 'What does Kuwait want to learn from TIMSS?' He replies in terms of whether students in Kuwait 'are taught the same curricula as the students in other countries' and whether they 'learn mathematics to a level that could be considered of a reasonably high international standard' (Hussein 1992, p. 468). Answering the same question, but with reference to the USA, Griffith and Medrich want to know what can be learnt 'from different educational structures and institutional practices that could support efforts [in the USA] to improve the performance of students and improve the quality of the schools' (Griffith and Medrich 1992, p. 482).

Apart from the obvious danger that international comparisons of student achievement run the risk of being burdened with responsibilities that they cannot realistically hope to meet, Orpwood's point, in general terms, is that it is essential to be clear about what international comparisons of student achievement in science and mathematics are meant to achieve. Are they principally intended to generate international 'league tables', i.e., rank orders of student performance on internationally deployed tests by country, region or system, or does the purpose go beyond this to embrace a study of factors such as teaching methods or curriculum that seem likely to influence the level of students' performance? If so, it becomes necessary to look beyond schools, teachers and students to take some account of how different communities 'define, deliver and organise instruction to provide educational opportunities to groups of students' (Schmidt and McKnight 1998). Whatever the answers to questions such as these, it seems clear that politicians and educational researchers in science and mathematics education are likely to continue to differ about the use they make of the results of international comparisons of student achievement. Ranking and league tables of performance will always have a greater appeal to politicians than to educational researchers, most of whom have little or no control of the use made of the results of their work.

A somewhat different limitation arises from the scale of the work and the organisation needed to ensure that it is undertaken effectively across a number of countries. This is not unrelated to Orpwood's concern about aims, since, the greater the number and diversity of the educational systems being studied, the more difficult it becomes to provide answers to the questions posed by policy makers, teachers and others.

Well-meaning attempts to cope with diverse systems will founder because of the difficulty of designing procedures and instruments which will have validity in all countries while still providing salient and significant data for those countries which share common aims, educational structures and contexts (Howson 1999, p. 167).

Issues to do with the scale and organisation of major international studies prompt a number of important questions about the direction, funding and control both of the research programme itself and of the analysis and dissemination of the results. Some of these questions have been explored by Keitel and Kilpatrick with respect to international comparisons of student achievement in mathematics, although their arguments are no less valid for parallel research in science. They point out that, in all the studies undertaken thus far by the International Association for the Evaluation of Educational Achievement, including TIMSS, almost all of the people with primary responsibility for conducting the study have been empirical researchers in education, psychometricians or experts in data processing. The problem of the content of mathematics or science education has been dealt with as no more than a technical question.

They argue that this has not only led to an emphasis upon methodological issues, such as reliability and validity, but to serious problems when one comes to the concrete aspect of mathematical tasks as representing curricula, teacher questionnaires as representing teaching, and results as yielding comparisons of curricula with respect to their cultural or social contexts (Keitel and Kilpatrick 1999, p. 245).

They further point out that 'the question of financial support influences whether the goals of the study are politically determined or are research oriented'. In the context of TIMSS, they come to the conclusion that 'a pseudo-consensus has been imposed (primarily by the English-speaking world) across systems so that curriculum can be taken as a constant rather than a variable, and so that the operation of other variables can be examined' (Keitel and Kilpatrick, *op.cit.*, p. 253).

A third group of limitations upon international comparisons of student achievement in science and mathematics might be described as technical or methodological in a narrower sense than that adopted by Keitel and Kilpatrick. They have to do with such issues as sampling and sample comparability, non-sampling errors and response rates, all of which ultimately influence the reliability and/or validity of the international comparisons undertaken. Even an issue such as defining the population of students whose performance is to be assessed presents significant problems.

Essentially, the choice is between an 'age-based' or a 'grade-based' sample population. Scrutiny of the results of any international comparison derived from a sample based on student age needs to recognise that children start formal schooling at different ages in different countries and that there is no uniform policy among countries about how frequently students are required to repeat one or more years of schooling before advancing to the next. A 'grade-based' sample, in contrast, encounters the difficulty that the content and sequence of a given grade in mathematics or science are not the same across educational systems. The TIMSS solution to this dilemma was to test students at two adjacent grades, but the consequences of this compromise solution are well-illustrated by Kitchen in this volume. Referring to the age distribution of children for the TIMSS Population 2, she notes that

> ... in addition to the class with most of the 13 year-olds, Scotland tested the class with mainly 12 year-olds and a few 13 year-olds while Japan tested the class with mainly 14 year-olds together with a few 13 year-olds ... [In France], only 80 per cent of the 13 year-olds were tested while some 20 per cent of the 15 year-old cohort were tested ... The missing 13 year-olds are those who have not yet reached the required standard while some of the 12 year-olds are those who have achieved the standard early and have moved to the next grade. There were even a few 11 year-olds.

The second qualification to be borne in mind when considering the usefulness of international comparisons of student achievement is that use does not 'necessarily mean supplying direct answer to questions, but rather in enabling planning and decision-making to be better informed' (Howson, 1999, p. 166). As Nisbet observed of research in science education, investigation 'sharpens thinking, directs attention to important issues, clarifies problems, encourages debate and the exchange of view, promotes flexibility and adaptation to changing demands'. In short, it aims to 'increase the problem-solving capacity of the educational system, rather than to provide final answers to questions' (Nisbet 1974, p. 106). Howson's reference to 'being better informed' can, of course, mean many things other than access to data that were not hitherto available or available only on a national or regional basis. It may, for example, involve understanding how other educational systems address and solve issues or problems that are common to most, perhaps all, countries (see Reynolds in this volume). It may also mean rendering problematic many aspects of the education process and the consequent construction of a substantial research and/or policy agenda relating, for example, to teacher training, classroom pedagogy and the use of the laboratory or of textbooks. Initially, therefore, a principal outcome of any international comparison of student achievement is the generation of a range of questions, the answers to which seek to account for any significant differences that have been established. To take an example from the 1995 TIMSS tests in England and Wales, why is it that the same groups of students in Population 1 and Population 2 performed less well in mathematics than in science, relative to their peers in other countries? Since the same groups of students were involved in taking the tests, the relatively poor levels of attainment in mathematics cannot be attributed to sampling errors or to wider social factors that influence schooling. The comparative levels of achievement of the same students in science and mathematics thus focuses attention on other issues, such as the form and content of the TIMSS tests and their relationship to the mathematics component of the national curriculum, national testing procedures, and what actually happens when science or mathematics is taught in laboratories and classrooms in England and Wales.

2.   EXPLOITING THE OUTCOMES OF INTERNATIONAL COMPARISONS OF STUDENT ACHIEVEMENT

It will be clear from the preceding paragraphs that considerable caution is needed in exploiting the outcomes of international comparison of student achievement in order to effect changes in educational policy and/or practice. Despite a range of obviously common elements, education systems are unique, serving and reflecting different communities, societies, regions and states. Just how unique is evident from the TIMSS encyclopaedia of *National Contexts for Mathematics and Science Education* (Robitaille 1997). It is thus difficult to sustain general conclusions from international comparisons of student achievement in science and mathematics, and probably impossible to offer any simple and speedy solutions to such problems as may be identified. As the Commissioner for Education Statistics in the USA advised the 1998 Conference of the National Science Teachers' Association, 'If you are looking for 'the reason' why our students performed as they did, you will be disappointed' (Forgione 1998, p. 19). What follows, therefore, are illustrations of some of the questions prompted by international comparative data on student achievement in science and/or mathematics, particularly those generated by the TIMSS initiative, together with some responses. These questions, of course, presuppose that the requisite degree of confidence can be placed in the international comparisons from which the questions are derived.

2.1   Shaping the curriculum

When the results of the second International Association for the Evaluation of Educational Achievement (IEA) study of student achievement became available, they showed that, while Korean primary students 'excelled in science achievement tests, Korean secondary students were falling behind their counterparts in science achievement'. This finding, according to Han, was 'immediately used in developing the 6th national curriculum, in which the content was reduced in order to allow increased emphasis to be placed on the development of students' basic science process skills' (Han 1995, p. 69). In Japan, data from the same international study indicated that Japanese students performed less well on tests of practical skills than on paper and pencil tests. The consequence was that courses of study were revised to 'emphasise observation and experiments, especially in the lower and upper secondary schools, and to attempt to develop and foster spontaneous inquiry activities and scientific thinking skills'. Likewise, evidence that Japanese students did poorly on science items with direct relevance to everyday life led to the view that practical subjects should be incorporated into the curriculum at the national level (Watanabe 1992, p. 458).

While not everyone would perhaps be quite so ready to make such firm and straightforward connections between elements of a prescribed curriculum and the relative level of students' performance, the assumption that there is a link between what (and how) students are taught and how well they perform in international tests of achievement is difficult to contest. Discussions about the nature of this link are an important and enduring consequence of the TIMSS programme which has often served to rejuvenate, and sometimes even to frame, academic and policy debates in mathematics and science education. As Howie notes later in this volume, the TIMSS results for South Africa 'not only focused the country's attention on the poor overall performance of South Africa's students' but 'revealed that the...Grade 7 and Grade 8 science and mathematics curricula were largely out of synch. with the other countries in the study'.

The TIMSS study has produced information about issues such as the range and number of topics included within science and mathematics curricula in many countries, together with data about the amount of time devoted to homework and to classroom/laboratory activities of various kinds. Consideration of the international TIMSS documentation alongside the individual national reports has thus prompted discussion about the form and content of science and mathematics curricula. In the USA, for example, mathematics curricula at all levels of schooling were found to contain far more topics than was the case in most other countries, a finding that may be directly related to the relatively poor performance of US students in the TIMSS mathematics tests. In science, the picture was somewhat more complicated, although science textbooks used in the USA also contained a larger number of topics than the international average. While it is important to acknowledge the variety of science courses provided in schools in the USA, the TIMSS science data also prompt the question of whether the number of topics taught is too large, especially in the middle years of schooling.

The number of topics taught is, of course, only a small part of the curriculum picture. How does the amount of time devoted to a given topic in one country compare with the international average and can any differences in time be related to differences in the levels of student achievement? Comparable questions can be asked about the amount of time devoted to homework in science or mathematics. Time, however, is a very crude dimension along which to undertake international curriculum comparisons since what matters is pedagogy, i.e., how effectively the time available is used to promote student learning. Is homework, for example, an integral part of the teaching and learning programme? If so, what kinds of homework are set and how and when is the work done at home followed up at school? The more important comparators, therefore, are likely to have to do with such matters as students' 'time on task', students' and teachers' assumptions about the standards to be achieved, and the teaching strategies deployed.

Different educational systems also sequence their science and mathematics curricula in different ways. In some instances, the science curriculum from the seventh grade onwards allows students to study only *one* of a number of sciences such as biology, chemistry or earth science, so that one year's study of biology might be followed a subsequent year studying chemistry. In other contexts, students study more than one of the scientific disciplines throughout part or all of the period of compulsory schooling. In England and Wales, for example, all pupils are required to follow a broad and balanced course of science from the age of five until the statutory school leaving age of 16. Again, it is important to try and establish the relationship, if any, between these different approaches to sequencing a curriculum and the students' performance in TIMSS tests.

Zuzovsky, in her contribution to this volume, shows how international comparative data upon student performance can be used, with the results of other studies, to explore the impact of curriculum change. Drawing upon curriculum reform in Israel, she notes that 'the Tomorrow 98 program does not seem to have advanced medium and even lower achievement ranks of elementary Israeli pupils in international comparisons'. She concludes that much more needs to be known about what is going on in science classrooms in that country.

2.2   Influencing pedagogy

One of the innovative aspects of the 1995 TIMSS programme was the attempt to investigate and characterise the instructional practices in six countries (France, Japan, Norway, Spain, Switzerland and the USA). The researchers coined the term 'characteristic pedagogical flow' (CPF) to describe the recurrent patterns of teaching and learning activities identified from a large number of classroom observations, supported by teacher questionnaires, in the various educational systems (Schmidt and McKnight 1995). Differences in pedagogical flow reflect differences in the ways in which mathematics or science is taught in different countries, and they derive from the interaction of teaching activities and curriculum within a context that is strongly shaped by the subtle social and cultural features of individual countries. The research revealed that different ways of engaging students with subject matter have evolved in different countries, these differences finding expression in, for example, the emphasis and focus of lessons, the kinds of explanation given to students, the attempts to relate abstract subject matter to students' everyday experience, the degree of responsibility given to students to develop and monitor their own learning, the cognitive demands made of seemingly comparable groups of students and the ways in which concepts are introduced, developed and linked to others.

Further international insights into pedagogy were provided by the TIMSS videotape study, although this was confined to an analysis of eighth grade mathematics classrooms in only three countries (Germany, Japan and the USA). The research again revealed differences in the mathematical content of lessons, in the ways in which mathematical ideas and procedures were presented, in the kinds of work expected of students, and in the extent and nature of classroom discourse (Stilger *et al.* 1998).

The fact that the teaching of mathematics and science, as with any other component of the school curriculum, is strongly socially, culturally and historically contingent is, of course, hardly a surprising finding, and the issues which it raises are central to the difficulties of conducting, and interpreting the outcomes of, international comparisons of student achievement. An important corollary is that significant elements of teachers' work are so contextually dependent and embedded that they are tacit and may be inaccessible to even the skilled observer. These elements cannot normally be transferred from one context to another and they cannot be justified by an appeal to the need to establish evidence-based practice. Not all aspects of teachers' work in classrooms and laboratories is, of course, of this kind and there is a temptation to try and transfer these other, more readily observable and explicated elements of teachers' work from one educational context to another, especially when they appear to be successful, as judged by student performance on international tests. While this is a temptation to be resisted, it is important to acknowledge that differences in 'pedagogical flow' within educational systems provide a valuable opportunity to review critically one's own practices in the light of the wider international picture. Mathematical or scientific concepts and procedures can be presented in a variety of ways, and different strategies are usually available to solve the problems commonly presented in school science or mathematics. Kaiser, drawing upon a qualitative comparative study of mathematics teaching at the lower secondary level in schools in Germany and England, reports a number of differences and identifies some possible consequences of the teaching of mathematics at this level in the two countries.

> Of high relevance is the different understanding of mathematical theory dominating English and German mathematics teaching ... the subject-oriented understanding of mathematical theory in German mathematics teaching has to integrate pragmatic elements – in contrast to the English mathematics teaching, which has to consider subject-oriented aspects. This implies detailed changes to mathematics teaching in both countries [including] the level of the teaching and learning styles ... [the] inclusion of individualised teaching and learning forms in German mathematics teaching and forms of class discussion in English teaching (Kaiser 1999, pp. 149-50)

Two further points about this international comparison of school mathematics teaching are of some interest. The first is that, despite major differences in scale, purpose and methodology, the ranking of student performance (in three countries) broadly mirrored that established by TIMSS. The second is that, as the project has expanded to include more countries, its distinctive feature, a qualitative, observation-based and longitudinal exploration of progression, appears to have been weakened (Howson 1999, p. 174).

It was noted at the beginning of this chapter that the best comparative studies draw, as appropriate, upon both qualitative and quantitative approaches. It is important, therefore, not to overlook the potential of the outcomes of the TIMSS tests as a source of information about what goes on when science or mathematics is taught in various educational systems. Angell and his colleagues, in this volume, show how responses to the free-response items in TIMSS can be used to shed light upon how students in a number of countries think about a range of scientific phenomena. The study offers many insights that are important for teachers and they conclude that students 'intuitive ideas do not need to be replaced so much as developed and refined'.

More generally, when the results of student performance are considered alongside those factors that determine 'pedagogical flow' within an educational system, it becomes possible to ask seminal questions about the relationships between pedagogy and the standards achieved by students in science and mathematics. Having reviewed several international studies, and concluded that the performance of English students in mathematics could not be regarded as 'anything other than poor', Reynolds and Farrell identified a number of factors that seemed to be influential in determining a high level of student achievement in mathematics in other countries (Reynolds and Farrell 1996). These were grouped as cultural (e.g. assumptions about learning, the status of teachers, high levels of commitment from parents and children), systemic (e.g., amount of time spent in school, a belief that *all* children are able to acquire core skills), school factors (e.g., the use of specialist teachers, frequent testing and monitoring), and classroom factors (e.g., the amount of whole-class interactive instruction, mechanisms to ensure that students who fall behind are given extra work to allow them to catch up). While acknowledging that it is a 'slightly risky enterprise, intellectually and practically' to look at non-English contexts and assess which of their practices may be helpful in raising the standards of work in English schools, they nonetheless conclude that 'the situation in which England finds itself is now so worrying that the risk involved in looking outward and trying new practices is worth taking'. They add that 'limited experimentation with non-British practice seems positively overdue' (*ibid.* 59). It is difficult to avoid the conclusion that thinking of this sort lies behind the centrally directed 'numeracy hour' introduced into primary schools in England in September 1999, although few teachers are likely to look upon this, with its detailed pedagogical advice, as a form of 'limited experimentation'.

A somewhat different approach to using the outcomes of TIMSS comparative studies is the TIMSS Resource Kit. According to the TIMSS web site, this is offered as a 'catalyst for careful contemplation, open discussion and considered action'. It is designed to be ... shared in groups ... among the education community, public decision-makers, community leaders, and the general public – to enlighten, explain and stimulate'. For Nebres, this Resource Kit constitutes an 'excellent starting point' for what he refers to as 'benchmarking', defined as the 'search for best practices that lead to superior performance' (Nebres 1999, p. 200)

2.3   Revisiting the tests

The public emphasis upon reporting of TIMSS data in the form of 'league tables' of the average performance of students in a number of countries disguises the wealth of detail which the study has provided about each of the countries participating in the study. Much can be gleaned from a close scrutiny of the performance of students in one country, especially when this is undertaken at the level of individual items or sub-set of items. Although it is necessary to continue to emphasise that the policy implications of such a scrutiny are rarely straightforward (Atkin and Black, 1997), the scope for secondary analysis of international data of the richness generated by TIMSS and other studies remains wide. The following provides some examples.

2.3.1   Students knowledge of science or mathematics

Consider the following question.
Air is made of many gases. Which gas is found in the greatest amount?
a) Nitrogen
b) Oxygen
c) Carbon Dioxide
d) Hydrogen
The international average for the correct response (a) to this question was 22 per cent at seventh grade and 27 per cent at eighth grade. In only five countries was the percentage of students choosing the correct response greater than 50 and none exceeded 60. In addition, the percentage of correct responses by students in a number of countries with advanced industrial economies was less than 10 per cent. The most common incorrect response in all cases was response (b) (oxygen). Given that the composition of the air is a feature of school science education common to all the countries in the TIMSS study, the responses to this test item can only be regarded as severely disappointing.

A similar lack of rudimentary knowledge was evident in a number of the mathematics items in the TIMSS tests. Asked which of $(m+4)$, $4m$, m^4, and $4(m+1)$ is equivalent to $m+m+m+m$ when m represents a positive number, the international average percentage of seventh and eighth grade students giving correct answers were 47.3 and 57.7 respectively. The corresponding percentages for the highest scoring country were 77.4 and 81.6 per cent respectively.

Howson uses the following item:

A person's heart is beating 72 times a minute. At this rate, about how many times does it beat in one hour?

a)   420,000
b)   42,000
c)   4,200
d)   420

to ask why 'after over eight years of formal education fewer than half' of [the] students in England gave the correct response (Howson, 1999, p. 185). He might have added that for the highest scoring country, 85.6 and 88.0 per cent of the students answered the question correctly.

Students' answers to questions offer valuable insights not simply into how well students are performing but also into their understanding of scientific or mathematical ideas. Data such as those generated by TIMSS make it possible to establish on a country basis, as well as internationally, what kinds of partially correct or incorrect answers students give to individual questions. In their chapter in this volume, Angell and his colleagues illustrate how students' answers to the free-response items within TIMSS can be coded and analysed, and the results used by teachers for diagnostic purposes. There is no reason why a parallel analysis could not be conducted using the responses to the fixed response items.

2.3.2   Progression in students' understanding of science and mathematics

The notion of student progression is central to schooling. It underpins the work of the National Assessment of Educational Progress (NAEP) in the USA and is embedded in numerous goals, attainment targets and performance levels that have come increasingly to characterise national, state or regional curriculum statements or guidelines. It is also fundamental to the attempts to define performance, especially in the vocational field, in terms of competencies and skills (Bates, 1995). This is despite the fact that, in most cases, progression is a poorly understood and inadequately conceptualised notion. Fensham, for example, has identified five different assumptions about the nature of progression in student learning in the school science curriculum projects of the 1960s and 1970s.

While shedding little light of a fundamental kind upon progression in student understanding, the TIMSS data generated at different grades allows changes in the average levels of student performance over time to be measured. The students are not, of course, the same students. What is being compared are samples of students drawn from each grade. Subject to this limitation, a direct comparison is possible because the tests used in the two sample TIMSS populations contained a number of common items.

TIMSS included fifteen mathematics questions which were common to the test booklets used with Population 1 and Population 2, although not all of these were in the published ('released') item pool. Using country-specific TIMSS data, it is possible to undertake a cross-age comparison and expose both the nature of any improvement in student performance and place any such improvement in the wider international context. In the case of England, for example, there was a predictable step-wise progression across the four year-groups, with a wider gap between the two Populations than between the two years within a Population. In addition, the performance by pupils in England on the 15 link items generally showed a greater degree of improvement over time than that indicated by the international average performance (Shorrocks-Taylor *et al.* 1998, p. 78). In the case of science, there were 17 link items, ten of which have been published. A predictable step-wise progression in student performance is again evident, but the pattern of student achievement was more homogenous for test items relating to biology than was with case with items testing student's knowledge and understanding of topics drawn from physics and chemistry (Shorrocks-Taylor *et al., op.cit.* p. 141).

In principle, it is also possible to compare student performance on TIMSS with the levels achieved in earlier international tests, e.g., the Second International Mathematics Study (SIMS) and the corresponding science study (SIMS). There were, for example, 22 link items between SIMS and TIMSS and country-specific analyses are possible. Using England as an example, Shorrocks-Taylor and her colleagues have shown that 'Performance on the link items did not reflect a generally perceived view that comparative performance on the TIMSS tests as a whole showed a poorer performance than previous studies'. They conclude that:

> Performance on the link items varied in both directions, but with a greater number of items showing a higher performance on the TIMSS tests than on the SIMS tests. The comparative decline in performance previously reported (Keys *et al.*, 1996) referred more specifically to the relative position of England in the international performance figures. This suggests that whilst performance in England has slightly improved, that of certain other countries may have improved to a greater degree. (Shorrocks-Taylor *et al., op.cit.*.p. 84)

The performance of students compared with their peers in other countries at the three grade levels tested within TIMSS may also shed some light on the relative progress of students as they pass through school. The 1999 TIMSS programme is important here since it will allow a comparison of the performance of a sample of those who were approximately nine years of age in 1995 (i.e., TIMSS Population 1) with that of a sample of those four years older (approximately 13 year-olds) in 1999. However, some insights can be gleaned from the existing TIMSS analysis. In the USA, for example, students' performance in science was significantly above the international average at the fourth grade level, the students doing better than 19 other countries and being outperformed by students from only one country, Korea. At the level of the eighth grade, US students still scored above the international average but they were now outperformed by students from nine other countries. In the twelfth grade general knowledge of science test, students in 11 of the 20 countries involved did better than their counterparts in the USA who also scored below the international average. Caution is, of course, necessary to avoid reading too much into these data, especially since they are derived from three different tests and three different sets of countries. Nonetheless, it is possible to see how international comparisons of student achievement may provide policy makers and others with some understanding of how the performance of students changes relative to that of comparable groups of students in other countries. In addition, when empirical evidence about a decline with increasing grade in the relative international standing of students in one country is placed alongside other evidence, e.g., that topics covered in one grade in most other countries are not covered until a later grade in the country concerned, then this is at least *prima facie* evidence of what has been called 'curriculum drift' (Forgione 1998).

An analysis of this kind is often particularly informative when student performance on various elements of the science or mathematics tests is disaggregated. In the case of the USA, for example, it is clear that the weakest area of student performance, i.e., student performance relative to that of students in other countries, at both fourth and eight grade is in the sub-set of TIMSS items concerned with physical science (fourth grade) and physics and chemistry (eighth grade). In England and Wales, students in Years 8 and 9 (TIMSS Population 2) scored *above* the international average in mathematics in the sub-sets of items concerned with data representation, analysis and probability but *below* the international means for fractions and number sense, geometry, algebra, measurement and proportionality. In Population 1 (Years 4 and 5), however, students in England and Wales scored *above* the international means in geometry, data representation, analysis and probability but *below* the international means for whole numbers, fractions and proportionality, measurement, estimation and number sense, and patterns, relations and functions (Keys, Harris and Fernandes, 1996; Harris, Keys and Fernandes, 1997). As always, the difficulty remains relating difference of this kind to the curriculum experienced by students in different countries.

2.3.3   Test format

The TIMSS test booklets contained items that sought to test a variety of topics and skills and the emphasis on particular skills and concepts varied from one test booklet to another. Different types of test item were used and each booklet contained both mathematics and science items. In addition, the clusters of questions used to construct the TIMSS tests were often located differently within the various booklets. For example, in Population 1, question cluster 'E' and question cluster 'G' were placed differently. Question cluster 'E' was at the beginning of the booklet 1B and third in booklet 3A. Similarly, cluster 'G' was the beginning in booklet 6A and third in booklet 5A.

Given this, it is likely that the TIMSS tests presented a number of challenges to many students. Testing student performance in science and mathematics in the same test requires students to switch their thinking between mathematics and science and between very different types of skills. In many countries, this would be an unfamiliar demand. There is also evidence that the position of a question in a TIMSS booklet has an effect on performance (Shorrocks-Taylor, *et al.,* 1998), questions being answered consistently less well if they appeared later, rather than earlier, in a test booklet. This, of course, may not have had any significant effect on performance on TIMSS tests as a whole because the order of the question cluster varied between test booklets. However, it is interesting to note that location may have had an effect on the overall level of performance of students on the open-response items which consistently appeared at the end of the TIMSS test booklets.

One of the dimensions of TIMSS that has perhaps not had the attention it deserves is the work on 'Performance Assessment' (Harmon *et al.,* 1997). Twenty-one and ten countries entered students for a set of 13 practical tasks at eighth and fourth grade respectively. The tasks, categorised as science, mathematics or combined mathematics and science, required students to investigate a range of practical problems and overall performance was the sum of a number of components, e.g., Using Scientific Procedures, and Applying Concept Knowledge. There was no obvious relationship between the relative ranking of the participating countries in these performance tests and in the more widely publicised written TIMSS tests. While this is perhaps to be expected since the two types of test are so different, it raises a number of important questions. For example, which kind of test is the more sympathetic to the kind of mathematics and science education that policy-makers, teachers and others should seek to promote? What is the nature of the relationship, if any, between students' performance at problem-solving tasks in science or mathematics and the time devoted to these activities in classrooms or laboratories or the emphasis given to them in textbooks? Any relationship is, of course, bound to be complex but, as with other outcomes of TIMSS, the data are rich and deserve careful scrutiny.

2.3.4   The role of language

Although opinions differ about the relationship between language and thought, few would doubt that the link is an intimate one. As Howie notes in this volume, South African students 'showed a lack of understanding of both mathematics and physical science questions and an inability to communicate their answers in instances where they did understand the questions'. Sjøberg has illustrated some of the difficulties of translating questions written in English into another language, using the following item included within the First International Science Study (FISS).

A certain wild bird has webbed feet. In which of the following places would you be most likely to find it?

a) a forest
b) a meadow
c) a cornfield
d) a desert
e) a lake

In the Swedish form of the question, 'webbed feet' is translated as *simhud* (swimming skin). This, in effect, reduces the task presented by the question to connecting swimming with a forest, a meadow, a cornfield, a desert or a lake. Given this, it is hardly surprising that 91 per cent of ten year-old Swedish students tested chose the correct answer. In Norway, where the word *svømmehud* is an apt translation, the problem was overcome in the Second Study (SISS) by showing a picture of a bird's foot. In Sweden, on the other hand, the question was replaced by another with a similar functional content, namely 'Where would you expect to find a bird with long legs and a peaked beak?' As Sjøberg observes, it 'does not take much imagination to see that all three items are rather different' (Sjøberg 1992, p. 4). The question reappears in diagrammatic form in TIMSS 1995.

The difficulties of seemingly straightforward translation should thus not be underestimated, nor should the fact that performance levels can vary depending upon whether information is presented visually or verbally. At worst, science or mathematics questions can be reduced largely, or even entirely, to questions of language. Readability of test items is also an issue, although this is not easy to estimate in the case of mathematics or science (see Shorrocks-Taylor *et al.,* 1998). The role of language in testing, however, involves much more than questions of translation or readability and more needs to be known about its relationship to student performance.

2.3.5 Gender Issues

Gender is an important variable in constructing test items and in analysing students' responses to them. In general, the mean scores of boys and girls were not statistically significant in most countries for either of the TIMSS Populations, although some statistically significant differences were identified involving, for example, Year 8/Population 2 students in England, Switzerland and Japan. Although statistically significant differences were few in number, they were always in favour of boys. Once again, however, a more detailed analysis of student response in terms of gender, item type and topic reveals the details of how an average test score is constructed for boys and girls. In the case of students in England, for example, there was no appreciable difference in the performance of boys and girls on multiple-choice items, as a reading of some of the literature might have predicted.

The gender differences in the responses to the TIMSS science tests were much more marked than in the case of mathematics. In Population 1, the majority of gender differences were in favour of boys, although, at both of the grades tested, these difference were statistically significant in fewer than half of the countries. In Population 2, there were significant gender differences in eleven countries when student responses were analysed in terms of the TIMSS reporting categories of earth science, physics and chemistry. Closer scrutiny of the performance of students in a given country allows attention to be given to the influence of item type and topic on the tests scores obtained by boys and girls respectively. It is also possible to explore whether any gender differential widens or narrows along either or both of these dimensions between the ages represented by the two TIMSS Populations.

3. CONCLUSION

Two things are perhaps clear from this chapter. The first is that well-conducted international comparisons of student achievement in science and mathematics, and TIMSS in particular, are a rich source of data about science and mathematics education in many countries. There is much that is yet to learned from the outcomes of these studies, especially when they are examined in fine detail and considered alongside other relevant findings, e.g., those relating to school improvement and effectiveness. The second is that international differences in student achievement in science or mathematics are much less important than a detailed understanding of what lies behind them, i.e., how they can be explained. Establishing such explanations is difficult. It requires extreme sensitivity to the social, cultural and historical contingency of education systems and what goes on within them. It is equally necessary to avoid confusing correlation with causality but that there is much to be gleaned seems incontestable.

REFERENCES

Atkin, J. M. and Black, P. J. (1997) Policy Perils of International Comparisons: The TIMSS Case, *Phi Delta Kappa*, 79, 1, pp. 22-8

Bates, I. (1995) The Competence Movement: Conceptualising Recent Research, *Studies in Science Education*, 25, pp. 39-68

Forgione, P. D. (1998) *What We've Learned From TIMSS About Science Education in the United States'*, Address to the 1998 Conference of the National Science Teachers' Association, http://nces.ed.gov/timss/report/97255-04.html

Griffith, J. E. and Medrich, E. A. (1992) What does the United States want to learn from international comparative studies in education? *Prospects*, 22, 4, pp. 476-85

Han, Jong-Ha. (1995) The Quest for National Standards in Science Education in Korea, *Studies in Science Education*, 26, pp. 59-71

Hans, N. *Comparative Education: A Study of Educational Factors and Traditions*, London

Harman, M., Smith T. A., Martin, M. O., Kelly, D. L., Beaton, A. E., Mullis, I. V. S., Gonzalez, E. J. and Orpwood, G. (1997) *Performance Assessment in the IEA's Third International Mathematics and Science Study*, (Chestnut Hill, MA: Boston College)

Harris, S., Keys, W. and Fernandes, C. (1997) *Third International Mathematics and Science Study, Second National Report, Part 1* (Slough: NFER

Howson, G. (1999) The Value of Comparative Studies, in G. Kaiser, E. Luna and I. Huntly, *op.cit*, pp. 165-88

Hussein, M. G. (1992) What does Kuwait want to learn from the Third International Mathematics and Science Study (TIMSS)? *Prospects*, 22, 4, pp. 463-8

Kaiser, G. (1999) Comparative Studies on Teaching Mathematics in England and Germany, in Kaiser, G., Luna, E. and Huntly, L. (eds.) *op.cit.*, pp. 140-50

Kaiser, G., Luna, E. and Huntly. I. (eds.) *International Comparisons in Mathematics Education* (Falmer: London)

Kandel, I. L. (1955) *The New Era in Education: A Comparative Study* (London: Harrap)

Keitel, C. and Kilpatrick, J. (1999) Rationality and Irrationality of International Comparisons, in G. Kaiser, E. Luna and I. Huntly (eds.) *International Comparisons in Mathematics Education* pp. 241-56 (Falmer: London)

Keys, W., Harris, S. and Fernandes, C. (1996) *Third International Mathematics and Science Study, Part 1* (Slough: NFER)

Nebres, B. F. (1999) International Benchmarking as a Way to Improve School Mathematics Achievement in the Era of Globalisation', in G. Kaiser, E. Luna and I. Huntly (eds.), *op.cit.*, pp. 200-12

Nesbit, J. (1974) Fifty Years of Research in Science Education, *Studies in Science Education*, 1, pp. 103-12

Noah, H. J. and Eckstein, M. A. (1969) *Towards a Science of Comparative Education*
   (Toronto: Macmillan)

Reynolds, D. and Farrell, S. (1996) *Worlds Apart? A Review of International Surveys of
   Educational Achievement involving England* (London: HMSO)

Robitaille, D. F. (ed.) (1997) *National Contexts for Mathematics and Science Education. An
   Encyclopaedia of the Education Systems Participating in TIMSS* (Vancouver: Pacific
   Educational Press)

Schmidt. W. H. and McKnight, C. C. (1998) So What Can we Really Learn From TIMSS?,
   http://TIMSS.msu.edu.whatcanwelearn.htm

Shorrocks-Taylor, D., Jenkins, E., Curry, J., Swinnerton, B., Hargreaves, M. and Nelson, N.
   (1998) *An Investigation of the Performance of English Pupils in the Third International
   Mathematics and Science Study (TIMSS)* (Leeds: Assessment and Evaluation Unit/ Centre
   for Studies in Science and Mathematics Education, University of Leeds)

Sjøberg, S. (1992) The IEA Science Study, SISS. Some Critical Points on Items and
   Questionnaires, Paper presented at the Fifteenth Comparative Education Society in Europe
   Conference, Dijon, June 27th-July 2nd 1992

Stilger, J. W., Gonzales, P., Kawanaka, T., Knoll, S. and Serrano, A. (1998) *Methods and
   Findings of the TIMSS Videotape Classroom Study* (Washington DC: US Government
   Printing Office)

Watanabe, R. (1992) How Japan Makes Use of International Educational Survey research,
   Prospects, 22, pp. 456-62

Chapter 8

Exploring Students Responses on Free-Response Science Items in TIMSS

Carl Angell, Marit Kjaernsli and Svein Lie
University of Oslo, Norway

INTRODUCTION

The scope and complexity of TIMSS (Third International Mathematics and Science Study) is enormous. Two monographs describe the different aspects of TIMSS in detail: the curriculum frameworks applied in the study (Robitaille *et al.*, 1993), the research questions and study design (Robitaille and Garden, 1996), and an overview is provided by Chapter 2 of this volume. Obviously, the main purpose of the TIMSS data has been to provide a basis for the construction of reliable achievement scales in the various content domains. Students are from three populations (Population 1 – 9 year-olds, Population 2 – 13 year-olds, and Population 3 – last year of secondary school), and the international science reports (Beaton *et al.*, 1996, Martin *et al.*, 1997, Mullis *et al.*, 1998) have reported between-country comparisons of averages (with standard errors) of student scores on these scales. In addition, comparisons between different sub-samples of students have been presented in a number of national reports and achievement scores have been related to a number of other variables such as students' home background and attitude towards the subjects, teachers' style of instruction, and school and class size.

There is no doubt that all these data provide indicators of the strong and weak aspects of school science, informing politicians and educators of necessary or possible steps that could be taken in structural and curricular reforms.

D. Shorrocks-Taylor and E.W. Jenkins (eds.), Learning from Others, 159–187.
© 2000 *Kluwer Academic Publishers. Printed in the Netherlands.*

However, there is another important aspect of the achievement data in TIMSS and similar projects. Achievement items are much more than just tools for constructing reporting scales. Data on any such item constitute a rich source of information, not only along the dimension of right/wrong (*How much* do they know?) but also on the diagnostic aspect as to which 'right' or 'wrong' responses (if any) students actually gave (*What* do they know?). The aim of this chapter is to draw more attention to this second aspect by presenting some item analyses from a science educator's point of view.

The TIMSS achievement tests included both multiple-choice and free-response (FR) items. We will try to demonstrate by using some examples that FR items do provide enriched insight into students' thinking, their conceptual understanding and the nature of their misconceptions. In particular this is true for FR items that require a more elaborate response in the form of explanations, justification or calculation.

Large-scale quantitative studies tend to be ignored or criticised by researchers in science and mathematics education. There seems to be a large gap between, on one hand, the statistical, psychometric testing approach and, on the other the currently popular qualitative subject matter-oriented point of view. Our position is that quantitative and qualitative approaches for probing students' thinking should be used together in a combined approach rather than in opposition. We will argue that coding and analysing FR items in TIMSS does represent a link between the two approaches.

A major strength of the TIMSS study is that students from many countries were tested over an extensive content area. For many of these areas, the analyses of responses can be linked to research on students' knowledge and understanding in science. Since the fundamental paper by Driver and Easley (1978) there has been a large number of studies of students' conceptions within a range of science topics. Many conferences have been held in this field (a series of three 'International Seminars on Misconceptions and Educational Strategies in Science and Mathematics', Cornell University, Ithaca, USA), and overviews (Wandersee *et al.,* 1993) and bibliographies (Pfundt and Duit, 1994) have been published. The theoretical paradigm for this large research activity is the so-called constructivist view of learning. The core of this theory is that students learn by constructing their own knowledge. When outer stimuli are treated in the mind together with earlier knowledge and experience of the issue, new insight is gained. Obviously, within such a framework, it is of crucial importance for teachers to be aware of the students' preconceptions within a topic prior to instruction in order that successful learning can occur. Therefore, the item-by-item (and even by country) results that are now available on the TIMSS home page (*http://wwwsteep.bc.edu/timss*) should be a rich and important source for researchers so that they can inform teachers, thereby improving science teaching world-wide.

2.   THE DEVELOPMENT OF CODING RUBRICS – THE TWO DIGIT SYSTEM

An imperative for making diagnostic quantitative analyses of responses to FR items is a coding system, which encompasses both the correctness dimension and the diagnostic aspect. In TIMSS this was provided by a two-digit system originally proposed by the Norwegian TIMSS team (Angell and Kobberstad, 1993, Angell *et al.*, 1994) and therefore sometimes referred to as the 'Viking Rubrics'. The Norwegian team also contributed substantially to the actual development of codes based on student responses in the field trial (Kjærnsli *et al.*, 1994, Angell, 1995). By applying this set of codes to the international field trial data, the Free Response Item Coding Committee (FRICC) developed the final set of codes (TIMSS 1995a, TIMSS 1995b). The process of development and the scope and principles of the coding rubrics have been further described by Lie *et al.*, (1996).

The fundamental basis of coding TIMSS FR items is *simplicity, authentic student-response orientation* and *acceptable inter-rater reliability.* For many items the correctness on one hand and method/error/type of explanation on the other are strongly inter-related. Instead of coding these two aspects separately, the idea behind the two-digit system is to apply only *one* two-digit variable that takes these issues into account.

The following *general* rubric illuminates the fundamental idea of the classification.

Code	Text	Score
20 - 29	Correct Response	2
10 - 19	Partial Response	1
70 - 79	Incorrect Response	0
90	Crossed out/erased, illegible, or impossible to interpret	0
99	Blank	0

The first digit gives information about the score. The second digit indicates the method used, type of explanation/examples given or type of error/misconception demonstrated. The score (the dimension of correctness) is thus linked to the other integrated aspects in such a way that the data can be analysed both for correctness and for diagnostic information.

9 used as a second digit represents a response that is classified as 'other' (except in 99, see below), whereas all other last digits each refer to a distinct category of responses, explicitly described in the coding guides (TIMSS 1995a, TIMSS 1995b).

The above distinction between codes 90 and 99 was made for the purpose of sorting out 'not reached' from 'reached, but not answered'. This distinction was essential for calculating item difficulties, but will play no role in the diagnostic analyses presented here. These two codes are, therefore, combined in the following discussion.

It was essential that response categories were developed on the basis of authentic student responses, and that the codes were constructed for each item independently. It was not an aim to construct universal categories based on theoretical considerations only. On the other hand, insight into the research on students' ways of thinking in many cases helped to focus some of the codes on well-known common misconceptions.

Another important feature is worth mentioning here. After the coding of the responses, and when analysing the data, the codes for a particular item could easily be combined in many different ways according to the aim of the analysis. This paper will show many examples of creating new categories out of a combination of original codes.

It should also be emphasised that a number of international training sessions were arranged to ensure reliable coding (Mullis and Smith, 1996). Further, during the process of coding, all countries were instructed to accomplish a within-country inter-rater reliability test. Sub-samples of approximately 10 per cent of the students' responses were coded independently by two raters. The percentage agreement was then calculated per item and per country. For Population 2, the average per cent agreement for science items were 95 per cent for the first digit (correctness score) and 87 per cent for both digits (exact agreement) (*ibid.*). For both the literacy and the physics test in Population 3, the reliability was almost exactly the same (Mullis *et al.*, 1998). A somewhat lower reliability was reported in a separate between-country reliability study, but this fact appears to be due to primarily situational and contextual differences in the way the measures were obtained. For instance, the coders from participating countries had to score responses in their non-native language (English), and a period of several months had passed since the scoring effort in their own countries (Mullis and Smith, 1996).

Finally, as a general feature of the coding system, we will report some statistics that show how the codes were actually applied and distributed. As an example, if we take the average for all science items and for all countries in the Population 3 literacy test, we can summarise as follows.

Out of all student responses given to the science FR items, there were:
- 28 per cent non-responses (code 90 or 99),
- 61 per cent responses within well-described categories (code with second digit 0, 1, 2, ... 6), and
- 11 per cent 'other' responses, e.g. responses other than those described by concrete categories (code 79, 19 or 29).

This means that the available data provides a detailed description of the great majority of responses, only around 10 per cent of the responses remaining unclassified by the applied coding rubrics. The results for the other populations are similar.

3. SOME SCIENCE LITERACY ITEMS

In the following we will show some examples of FR items from the science literacy test in Population 3, with their coding guides and results. Although the exemplary items mainly reflect the more content-based part of the literacy test, this test also contains some more contextualised items which are designed to measure the so-called 'Reasoning and Social Utility' (RSU) aspect. The various aspects of, and the rationale for, the mathematics and science literacy Population 3 test are thoroughly described in a special monograph (Orpwood and Garden, 1998, see also Orpwood's chapter in this volume).

High heeled shoes

A7
Some high heeled shoes are claimed to damage floors. The base diameter of these very high heels is about 0.5 cm. Briefly explain why the very high heels may cause damage to floors.

Item A7 assesses students' understanding of the physical concept of pressure in a daily-life context. Do they understand that pressure will be greater if the area of the heels get smaller? And how can they formulate their understanding with or without the relevant scientific terminology?

It is remarkable how many students have given an answer to this item. Internationally, 87 per cent of the students have answered, and in Norway as many as 95 per cent. Further, many of them have demonstrated at least a partial understanding of the concepts involved.

Full score on this item gives 2 points. Table 8.1 shows the coding guide with the international distribution of responses. Two different categories of answers give two points. Here we have deleted code 29 because the international results show that almost no one (0.4 per cent) got this code. To obtain code 20 the student needs to refer explicitly to 'greater pressure' and to give an explanation (smaller area). It is remarkable how few students have answered correctly using an appropriate physical science vocabulary.

For code 21 the response does not use the concept of 'pressure', but rather concepts such as weight and force, and how these act on a smaller area. There was a discussion whether answers of this kind (not referring to 'pressure') really deserved two points or not. In the Norwegian data the students in category 21 had a lower overall score than those in codes 12 or 13. On the other hand it was also argued that code 10 should have been given 2 points. These students gave correct answers, even if they did not explain. The question here is whether we really ask for an expanded explanation of the answers in this item (*'briefly explain ...'*). On the other hand these students tend to have a relatively low overall score.

The flexibility of the codes implies that they allow analyses to be carried out also across the admittedly somewhat arbitrary 'correctness' dimension for items like this. For example, it can be inferred that, internationally, about 22 per cent of students correctly use the word 'pressure' (codes 20/10) whereas around 5 per cent of them incorrectly use the word as a synonym for force (code 12).

In Figure 8.1, we have combined codes 11, 12 and 13 for practical reasons. They all tell us that the students mix or misuse some of the words 'force', 'pressure', 'mass' or 'weight'. In spite of this, however, they may well have the correct idea. All these words are often used in our daily language in a less precise manner. The students may have a practical understanding based on experience, but when they try to use scientific vocabulary, they fail to apply it correctly. Or can it really be regarded as wrong as long as most people use it in that way? From a linguistic point of view it could well be argued that these students apply the terms 'correctly', i.e. according to common use in everyday language. The scientists do not 'own' the words. Let us show three examples, one for each of the three codes: *'The pressure is distributed over a smaller area.'* (code 12), *'The force increases as the area of the heel gets smaller.'* (code 11), and *'The mass is distributed over a smaller area.'* (code 13). All three responses reveal a correct thinking regardless of the 'incorrect' use of scientific terms. None of these three codes stands out. However, there are more students that misuse 'pressure' instead of 'force' than the other alternatives.

Even for code 70, one still might argue that the students have a partial understanding of the relevant phenomenon.

Table 8.1. Item A7, High heeled shoes: Coding guide and international results

Code	Response	International results (%)
Correct Response		
20	Refers to greater pressure on the floor because of smaller area of the heels.	19
21	Refers to weight or force acting on a smaller area or heel size, without using the term pressure.	22
Partial Response		
10	Refers to greater pressure without mentioning area of the heels.	3
11	Refers to an increased 'force' instead of 'pressure' with smaller area.	3
12	Refers to 'pressure' instead of 'force', but correct thinking.	5
13	Refers to 'mass' instead of 'force' or 'weight', but correct thinking.	3
19	Other partial	7
Incorrect Response		
70	Refers only to the hardness of the material or sharpness of high heels.	11
76+79	Repeats information in the stem / Other incorrect	16
Non-response		
90+99	Crossed out etc./ Blank	13

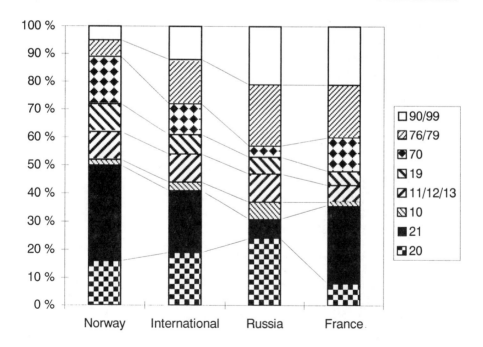

Figure 8.1. Item A7, High heeled shoes: International and some national results

In Figure 8.1 we have shown the international average results and the results of Norway, France and Russia. We have chosen Russia and France to show two countries that have almost the same proportion of correct answers. However, there is a big difference in the words and concepts the students have used. In Russia most of the students who got two points used the concept of 'pressure' explicitly, whereas in France it is the other way around. In Norway the tendency is the same as in France.

From Figure 8.1 we also see a difference between the countries in regard to the use of code 70 (answers such as *'they are sharper and they poke into the floor'*). In Norway a large majority of the incorrect responses received this code. An equal tendency is evident in France, but in Russia there are only 4 per cent such responses.

3.2   Thirsty on a hot day

| **B13** |
| Write down the reason why we get thirsty on a hot day and have to drink a lot. |

This item, being a 'link item', is very easy for Population 3 as the same item has been given also to both Populations 1 and 2. In all the participating countries more than 80 per cent of students received a full score on this item. Full score is one point. The coding guide is shown in Table 8.2.

Table 8.2. Item B13, Thirsty on a hot day: Coding guide and international results

Code	Response	International results (%)
Correct Response		
10	Refers to perspiration, its cooling effect, and replacement of lost water	15
11	Refers to perspiration and replacement of lost water	43
12	Refers to perspiration and its cooling effect	2
13	Refers to perspiration only	20
19	Other acceptable explanation	2
Incorrect Response		
70	Refers to body temperature (being too hot) but does not answer why we get thirsty	1
71	Refers only to drying of the body	4
72	Refers to getting more energy by drinking more water	1
76+79	Repeats information in the stem / Other incorrect	5
Non response		
90+99	Crossed out etc./ Blank	7

This item is an example of one in which the 'correctness score agreement' was about average for science (86 per cent internationally and 95 per cent within countries), but where 'diagnostic code agreement' was very low (59 per cent internationally and 80 per cent within countries) (Mullis and Smith, 1996). The reason for this is quite easy to understand. As one can see in Table 8.2, to get code 10 the students had to refer to perspiration and its cooling effect and the need to replace lost water. Code 12 was almost the same, but it was not necessary to refer to the replacement of lost water. When the students say that one sweats, they may be thinking that it is obvious and thus unnecessary to state explicitly that one has to replace water. Possibly, the terms used in different countries do not carry the same meaning in this respect, and may be different from language to language. Internationally just 2 per cent of the students got code 12, so in the further discussion and in Figure 8.2 we have combined the codes 10 and 12, since the most important issue here is to distinguish between students who do and those who do not refer to the cooling effect. We have the same problem with codes 11 and 13. Students had to refer explicitly to the need of replacing lost water in code 11, but not in code 13.

Given the above, it makes much sense to combine codes 10 and 12, and also codes 11 and 13. By doing so we also would increase the above mentioned diagnostic code agreement. As many as 14 per cent of the international disagreements for this item were 10-12 or 11-13 'disagreements' (Mullis and Smith, 1996, app. H). If we do not count these, the international diagnostic code agreement increases from 59 per cent to 73 per cent for the item.

It was discussed in the FRICC committee (see above) whether codes 10 and 12 represent better answers than the others and therefore 'deserve' 2 points. From a psychometric point of view this can be supported by the Norwegian data, as students in both these categories have much higher overall scores than all other categories of students. However, a closer look at the item itself reveals that the students are asked to write down *the reason why* we get thirsty, thus implicitly asking for *the one* reason, which obviously is sweating. A response which also refers to the *function* of sweating, namely temperature regulation of the human body, is definitely a more advanced response, but it cannot reasonably be given a higher score. This would have been different if the question had been phrased as *Explain why* ... This point illustrates the necessary close relation that must exist between the score points allocated and the exact phrasing of the question asked.

However, this example also gives another demonstration of the power and flexibility of the coding system. When performing a diagnostic analysis, the codes can be compared and combined according to the main issues under consideration. In the further analysis, the combined codes 10/12 are regarded as a more advanced response, thus contributing to more levels on an 'achievement scale' for the item.

For Population 3 the incorrect responses are not so interesting, because of the high degree of correctness. We will discuss this later in connection with Figure 8.3.

In Figure 8.2 we have displayed the international results and results for Norway, Australia and Russia. The Russian data for this item have a somewhat different profile than in the former item showed in Figure 8.1. In the present case the more advanced responses (10/12) are rather thinly represented in this country's data.

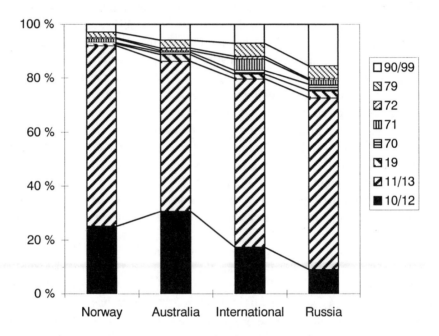

Figure 8.2. Item B13, Thirsty on a hot day: International and some national results

As noted above, this item is a link item, used with all three populations. Figure 8.3 shows the Norwegian results from all three Populations. The per cent of non-responses naturally decline substantially from Grade 2 to the last year of schooling. The most interesting feature of this comparison is probably how references to the cooling effect (discussed above) dramatically increase in frequency from Population 2 to 3 and from the vocational to the academic track of upper secondary school. Similarly, one can see how the wrong responses gradually disappear with age. But in the lower grades some interesting misconceptions are revealed.

In grades 2, 3 and 6 we see that some of the students state that they cool down the body temperature by drinking something *cold*. Even if formally true from a physicist's point of view, this effect is of vanishing importance (and therefore regarded as not correct) compared to getting enough liquid. In fact, it is easier to drink a lot when the liquid is not too cold.

Some of the students refer only to drying of the body, code 71. Examples are *'Your throat gets dry.'* and *'You get drier.'* Again, one can argue that such responses are 'correct' and deserve a score point, but it was judged as too simplistic. Interestingly enough, in France as much as 22 per cent of the Population 3 responses was classified as code 71.

Code 72 represents an interesting misconception. These students demonstrate the belief that you have to drink because you get exhausted, or that you get more energy by drinking water. This misconception is much more common than seen from the frequency distribution, simply because a number of responses mentioned sweating *in addition to* this erroneous statement and were therefore judged as correct.

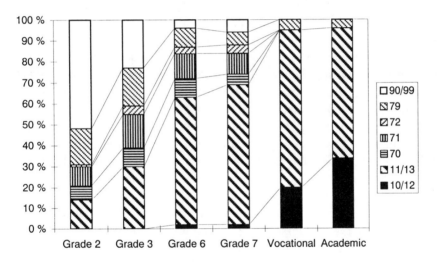

Figure 8.3. Item B13, Thirsty on a hot day: Norwegian results for all three populations

3.3   Kettle of boiling water

C20

A kettle of boiling water is on a stove. If the burner under the kettle is turned up, what will happen to the temperature of the water in the kettle? Explain your answer.

Item C20 probes students' understanding of heat and temperature, in the context of the fixed temperature of a boiling liquid. Simply stated, the item measures if the students know that the temperature of the boiling water is constant even if you add more energy. This issue has obvious practical implications for daily life.

Full score on this item is 2 points, and to get full score it is necessary to have to state that the temperature stays the same and to give a 'good' explanation for it, see Table 8.3. The best scientific explanation (code 21) refers to the fact that *'The energy will only change the intermolecular state'*, or *'The heat supplied by the burner will be used to evaporate the water'* or similar responses. Very few gave such an academic response, however. In the international results, just 2 per cent of the students got this code. In our further analysis later we have therefore combined codes 20, 21 and 29, as in Figure 8.4.

There is quite a distance in subject content between the answers coded 20 and 21. Responses within code 20 refer to the concept of a boiling point. However, one may well argue that such a reference cannot, from a strictly logical point of view, constitute any *explanation* of the phenomenon, but is rather a restatement of what is already said (constant temperature) either with or without using the scientific concept of boiling point. Two examples of actual code 20 responses are *'The temperature of the boiling water will always stay the same.'* and *'Once water is boiling, the temperature cannot reach a higher boiling point.'* In neither of these cases can we find any real explanation. This is an example of much more general issue in science. What constitutes a valid explanation is a judgement call and thus very context dependent. Decisions cannot be based on scientific or logical arguments alone.

Answers that fit into code 10 have mentioned that the temperature stays the same, so for any practical purpose this might be 'correct enough'. Possibly, these students do understand that the boiling point is 100°C. This fact might well be regarded as implicit in their answers.

The most interesting feature of this particular item is probably the students' misconception. Internationally, as many as 23 per cent of the students express the idea that the temperature continues to rise after boiling if you turn up the burner (codes 71, 72 or 73), see Figure 8.4. In the USA more than 40 per cent of the students gave such a response (23 per cent are coded as 71, 10 per cent as code 72 and 9 per cent as 73). If this misconception is turned into practice it can obviously lead to the use of more energy/gas than necessary, e.g. if more heat is added in order that potatoes or rice should be cooked faster after boiling has started.

It is worth mentioning that in a parallel Population 2 item (Year 2), students were asked about the temperature in a snowball after holding it in your hand for a minute.   Not surprisingly, as many as 24 per cent of the students stated that the temperature would increase.

Table 8.3. Item C20, Kettle of boiling water: Coding guide and international results

Code	Response	International results (%)
Correct Response		
20	Temperature stays the same: refers to 'boiling point' or 100 C or (increased) evaporation without explicitly mentioning energy or heat.	23
21	As in code 20, but refers to energy or heat explicitly	3
29	Other correct	2
Partial Response		
10	Temperature stays the same; refers only to more violent boiling	2
11	Temperature stays the same; explanation missing or incorrect.	3
19	Other partial	2
Incorrect Response		
70	Temperature not mentioned; refers to more violent boiling and/or more evaporation (steam).	6
71	Temperature will rise; refers to increased temperature of the burner and /or more energy or heat added.	10
72	Temperature will rise; refers to more violent boiling and/or producing more evaporation (steam).	6
73	Temperature will rise; no explanation.	6
76+79	Repeats information in the stem / Other incorrect	6
Non-response		
90+99	Crossed out etc./ Blank	30

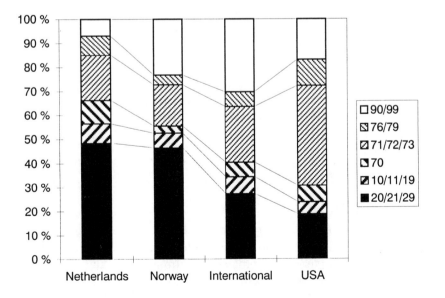

Figure 8.4. Item C20, Kettle of boiling water: International and some national results

Figure 8.4 shows the distribution of responses for some and for all countries. As mentioned above, we have here combined 20, 21 and 29 because of the scarce occurrence of codes 21 and 29. We have also combined codes 71, 72, and 73 because they all in different ways refer to students who believe that the temperature will rise. The possible waste-of-energy aspect of this response category should be a concern, particularly in the USA.

3.4 Rain from another place

> **C19**
>
> Draw a diagram to show how the water that falls as rain in one place may come from another place that is far away.

To get the top score on this item, it is necessary to display all three aspects of the water cycle (evaporation, transportation and precipitation) on a drawing, see Table 8.4. The good results from Norway and the Netherlands show that an understanding of this topic is considered important in school and in daily life, see Figure 8.5. The results can also be seen in the context of the greater annual precipitation in Norway and the Netherlands. In countries like Cyprus and Israel it seems that this is not an important topic. It is remarkable that such a large percentage of students from these two countries did not even respond to this item that tests a rather fundamental issue.

Table 8.4. Item C19, Rain from another place: Coding guide

Code	Response
Correct Response	
20	Response includes the three following steps:
	i. Evaporation of water from source.
	ii. Transportation of water as vapour/clouds to another place.
	iii. Precipitation in other places.
Partial Response	
10	As in code 20, but response does not include evaporation
11	As in code 20, but response does not include transportation
	As in code 20, but response does not include precipitation
19	Other partial
Incorrect Response	
70	Response indicates precipitation only; it may use vertical or diagonal lines.
79	Other incorrect
Non-response	
90+99	Crossed out etc./ Blank

An interesting feature emerged from the process of coding Norwegian responses. Some students who got no credit for evaporation (code 10) had made drawings that showed clouds coming from (often English) factories. By looking closer at these drawings it seemed that these students believed that the smoke from factory chimneys was the generator of clouds. This misconception could well stem from teaching the concept of acid rain. We uncovered this information during the process of coding.   There is no separate code for this misconception.

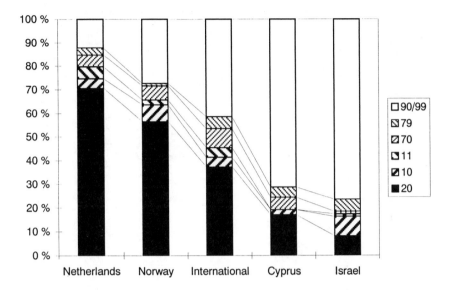

Figure 8.5. Item C19, Rain from another place: International and some national results

Figure 8.6. Item C19, Rain from another place: A student response (code 10)

Some of the students included precipitation only. Also, many of these drawings had diagonal lines indicating that rain was transported by wind over great distances (see Figure 8.7).

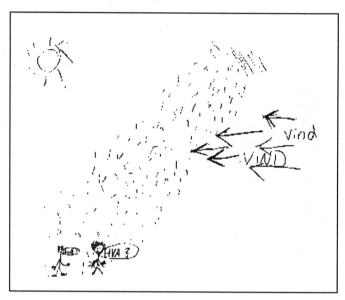

Figure 8.7. Item C19, Rain from another place: A student response (code 70)

4.   THE PHYSICS SPECIALIST TEST

Physics achievement results for students specialising in physics are reported for 16 countries in the TIMSS study. The percentage of the entire school-leaving age cohort that participated in the physics study was approximately 15 per cent in several countries, although it varied from 2 per cent in Russia to 39 per cent in Slovenia. In Norway the percentage was 8 per cent and in Sweden 16 per cent. Norway and Sweden had average physics achievement scores similar to each other but significantly higher than the other participating countries.

The physics items in TIMSS are about fundamental laws and principles which were supposed to be typical of physics courses at this level in schools. Most of the items are dealing with *one* central problem and they are less concerned with contextualised or everyday problems. This fact might well be criticised, but in our view this is also the strength of many of the TIMSS physics items. They are well suited for diagnostic analysis of students' fundamental understanding of physics.

The following examples are presented in order to show the benefit of the coding system and its potential for understanding and exploring student thinking. In addition, some specific physics problems connected with the selected items are discussed.

4.1   Acceleration arrows of a bouncing ball

Newton's laws of force and motion represent a topic within physics that is taught at many levels in schools around the world. These laws are apparently simple, at least the mathematical formulation of the second law, $F = ma$, seems to be simple. But it is not! All the concepts involved, force, mass and acceleration, are complicated and difficult to understand. Few students get a deep understanding of Newton's second law by calculating one unknown quantity from two known ones. A more qualitative approach is necessary to form a foundation for understanding the concepts involved. As we will show, even the so-called 'physics specialists' in many countries have great problems with some of the central and basic concepts. The arguably most fundamental law in mechanics (or even in physics) is simply not understood, and this should in our view be a matter of serious concern within science (physics) education.

G15

The figure shows the trajectory of a ball bouncing on a floor, with negligible air resistance.

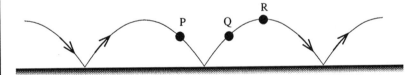

Draw arrows on the figure showing the direction of acceleration of the ball at points P, Q and R.

The problem in this item is well known from a number of research studies (e.g. Viennot, 1979; Sjøberg and Lie, 1981; Finegold and Gorsky, 1991; Ebison 1993; Wandersee *et al.*, 1993), but it should be noticed that most of these studies focused on which *forces* are acting and not on the *acceleration*. Such research studies have revealed a very common misconception referred to as 'impetus' or 'Aristotelian' ideas. Impetus is a historical idea about 'a moving force within the body' which pulls the body along the path after it has been thrown. 'Aristotelian' ideas refer to the 'law of motion' by Aristotle. Here a force is also needed to maintain motion, the force acting in the direction of the motion, and force and motion are proportional to each other.

However, when a ball is bouncing on a floor and we can neglect the air resistance as described, the acceleration is always pointing vertically downwards as long as the ball is not in contact with the floor. The only force acting on the ball is the gravity pointing downwards, and due to Newton's second law, the acceleration and the sum of the forces have the same direction.

The following coding rubrics show the actual codes for this item and the result for the international average as per cent.

Table 8.5. Item G15, Acceleration of a bouncing ball: Coding guide

Code	Response	Int. Average
Correct Response		
10	The acceleration is parallel to g, downwards at P, Q and R	16
Incorrect Response		
70	The acceleration is parallel to g, downwards arrow at P, upwards at Q and zero at R	7
71	The acceleration is parallel to g, downwards arrow at P, upwards at Q and either upwards or downwards at R	4
72	The acceleration has the same direction as the motion (at least in P and Q). Any response at R.	34
Code	Response	Int. Average
73	The acceleration has the same direction as the motion at P, the opposite direction from the motion at Q. Any response at R.	6
74	The acceleration has the direction perpendicular to the motion (at least at P and Q)	5
79	Other incorrect responses	21
Non-response		
90/99		7

The results show that this item is very demanding for students in many countries. An overall average of 16 per cent for correct response is rather low. There are considerable differences between countries, correct answers varying from 4 per cent to 46 per cent of the students.

In many countries, the students' answers indicate alternative conceptions (intuitive ideas) in two different ways or a combination of these. The acceleration has always the same direction as the motion (i.e. parallel to the velocity), and the acceleration is pointing upwards when, for example, a ball is moving upwards in a throw.

Probably code 70 describes the most specific response. The acceleration is parallel to g, downwards arrow at P, upwards at Q and zero at R. But only Sweden has a high percentage of responses which are coded 70 (24 per cent).

The most distinctive result is, however, the high percentage for code 72 which includes two misconceptions. The acceleration is parallel to the motion and the acceleration is pointing upwards when the ball is moving upwards. It becomes even more clear if we put some codes together. All the codes 70, 71 and 72 include the misconception that the acceleration points upwards when the ball is moving upwards. Internationally, an average of 45 per cent of the students give answers involving this misconception.

Another focus is revealed if codes 70 and 71 are combined as well as codes 72 and 73. Both codes 70 and 71 describe the acceleration parallel to *g*, but these codes include the misconception that the acceleration is upwards when the ball is moving upwards. Codes 72 and 73 both include the conception of acceleration parallel to the motion.

The following diagram shows results from some selected countries in order to illustrate the variation between countries.

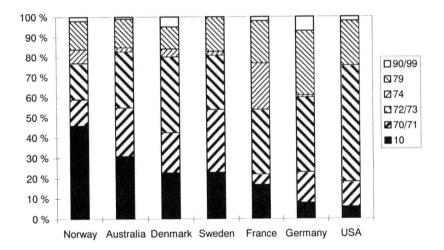

Figure 8.8. Item G15, Acceleration of bouncing ball: Results from some selected countries

Code 10:	*Correct*
Code 70/71:	*Acceleration is parallel to g, downwards at P and upwards at Q*
Code 72/73:	*Acceleration is parallel to the motion*
Code 74:	*Acceleration is perpendicular to the motion*
Code 79:	*Other incorrect responses*
Code 90 / 99:	*Non-response*

This result is astonishing. Even Sweden, with very good overall results, has a remarkably low percentage of correct responses to this item. In particular, code 70/71 is often used for the Swedish responses. In the USA, Germany and Denmark the use of code 72/73 is also remarkable. Code 74 is of little use in most of the countries. Only France is different. Code 74 describes the acceleration as perpendicular to the motion as if there is a circular motion.

As mentioned, many earlier studies have dealt with the concept of force and not acceleration. After the TIMSS study, we did a small survey in Norway just to investigate this difference. On a free-response item, very similar to the TIMSS item, we asked a sample of students to draw arrows showing the *force* acting on the ball. As many as 71 per cent of the Norwegian students drew correct force arrows (downwards in all cases), but only 46 per cent drew correct acceleration arrows in TIMSS. Even if we cannot compare these results directly, they provide some indication that the understanding of the kinematics (about movement) is different from the dynamics (about force). It seems that students have greater difficulties with the concept of acceleration than the concept of force when it comes to understanding the direction of these two quantities. An interpretation of this result is that the understanding of the vector aspect is easier for a force than for acceleration. This can reasonably be an explanation of the remarkable international result in TIMSS, even in countries where the vector aspect is emphasised in instruction.

Figure 8.9 shows the results of an extended analysis of the Norwegian data. The students are categorised in three scoring groups. Group 1 is the 25 per cent lowest achieving students measured at the total score scale. Group 2 is the 50 per cent in the middle and group 3 is the 25 per cent best achieving students. Also, in this analysis, codes 70 and 71 are combined, as well as 72 and 73.

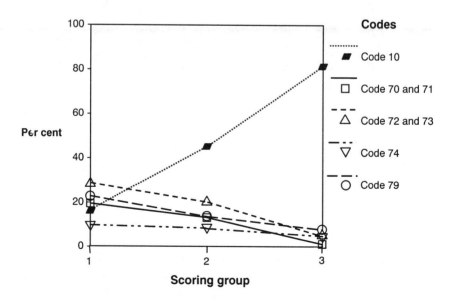

Figure 8.9. Item G15, Acceleration of bouncing ball: Norwegian results

10:	*Correct*
70 and 71:	*Acceleration points downwards at P and upwards at Q*
72 and 73:	*Acceleration has the same or opposite direction as the motion (parallel to velocity)*
74:	*Acceleration is perpendicular to the motion*
79:	*Other incorrect responses*

In spite of the good result in Norway compared with many other countries, it is important to emphasise that it was only among the best students (group 3) that a majority answered correctly. In addition, the middle achieving Norwegian students have a correct response frequency above the international average in TIMSS on this item. The most frequent type of non-correct responses students are those with the acceleration parallel to the motion (velocity). This is so for both scoring groups 1 and 2.

In Norway the gender difference is considerable for this item. Figure 8.10 shows the distribution of responses for girls and boys. There are significantly more responses from girls indicating misconceptions than for the boys. However, it is interesting to notice that the relative distribution of wrong responses is about the same.

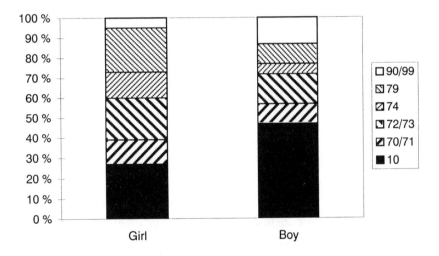

Figure 8.10. Item G15, Acceleration of bouncing ball: Sex differences in Norway

In Norway, as in many other countries, much research in science education has focused on students' conceptions of force and motion. For example, the Aristotelian concept of force and the impetus theory should be well known among physics teachers around the world. In Norway there has been a special attention to students' conceptions of force and motion for many years. This issue has been emphasised in textbooks and in teacher education and in-service courses for teachers. But in spite of this, students all over the world seem to a large extent to have the same ideas as before this emphasis was placed. From this perspective the large effort in revealing students' alternative conceptions is discouraging. This should be of serious concern for the community of science educators.

4.2   Falling ring and magnet

Unlike the field of mechanics, students' understanding of electro-magnetism is an area in which little research has been published. There have, however, been published a large number of articles about elementary electricity and electric circuits (Pfundt and Duit, 1994).

G19

A strong bar magnet hangs from a string with its north pole upwards. A light ring of aluminium is held above the magnet and allowed to fall down to the ground, as shown in the figure.

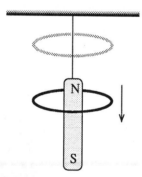

Explain why the ring takes longer to fall to the ground with the magnet present than it would without the magnet

---

This item is difficult, but it focuses on very fundamental ideas within electromagnetism. The magnetic field through the ring is changing while the ring is falling. Therefore there will be an induced current in the ring and this current produces a force acting on the ring to oppose the movement. This upward force will in turn cause a decreased acceleration and thus cause the ring to take longer to fall to the ground. The main point of this task is the force produced between the ring of aluminium and the magnet due to electromagnetic induction. There is *not* any form of 'direct' magnetic [8.1] force between the ring and the magnet. It should be a well-known fact that aluminium is a non-magnetic material.

[8.1] By a 'direct magnetic' force we mean what usually is considered as 'magnetism': a magnetic force between two magnets or forces between a magnet and a magnetic material such as iron.

As a matter of fact, this phenomenon can be very nicely demonstrated with modern computer based equipment.

Table 8.6. Item G19, Falling ring and magnet: Coding guide

Code	Response	International Average
Correct Response		
20/21	Responses refers to induction and a force acting on the ring in the opposite direction of the motion	13
29	Other acceptable responses such as reasons including conservation of energy.	1
Partial response		
10	Incomplete response, but refers to induction or Lenz's law	3
19	Other partially correct responses	4
Incorrect Response		
70	Responses expressing the idea that the magnet pushes (or pulls) on the ring due to the magnetic force from the magnet. Nothing recorded about induction	51
79	Other incorrect responses	13
Non-response		
90/99		15

Table 8.6 is a revised version of the coding scheme used for this item. As seen from the Table, the international average correct response on this item is low. Only 21 per cent of the students got one or two points, and as many as 51 per cent of the responses are coded 70. In some countries it seems that the problem stated in the task is not only unusual, but even incomprehensible for most of the students. In the USA, Canada, Austria and Czech Republic about 70 per cent of the responses are coded 70! It seems that induction is almost an unknown phenomenon among 'physics specialists' in many countries. As a matter of fact, Norway is the highest scoring country, and only Germany and Cyprus have results comparable to Norway. Figure 8.11 shows the results from some selected countries.

Figure 8.11 confirms the very poor result and the fact that code 70 is dominant in many countries. The students in this category have expressed the 'direct magnetic' idea that the magnet pushes or pulls on the ring due to magnetic forces without any reference to induction. In other words, there are many students who take no account of induction which is the fundamental phenomenon on which the question is based.

Internationally, there were almost no correct responses with explanations focusing on the conservation of energy. In the actual situation there is energy transformation from potential to kinetic energy, and then to electric energy and heat. This implies that the ring gets less kinetic energy and is therefore slowing down. It should be something to reflect on for physics teachers around the world that energy conservation, even well known in principle, is not applied in a situation like this. For the present task, reasoning with energy should be easier than complicated explanations with rules for direction of the current and the force. The concept of energy seems to be applied less often to problems like this than to problems in mechanics.

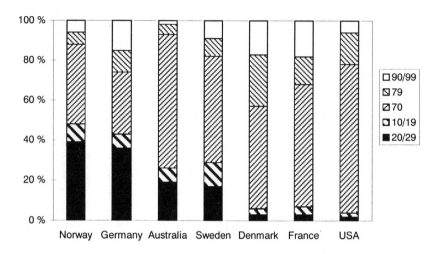

Figure 8.11. Item G19, Falling ring and magnet: results for some selected countries

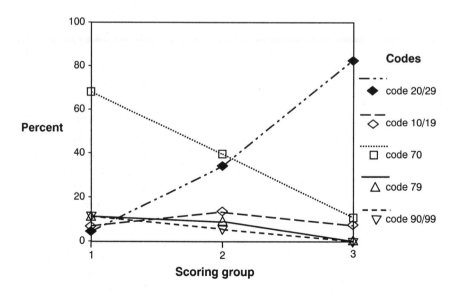

Figure 8.12. Item G19, Falling ring and magnet: Norwegian results

20/21:	Correct response, refers to induction and forces
10/19:	Partial response
70:	Responses refer to magnetism, nothing about induction
79:	Other incorrect responses
90/99:	Non-response

Figure 8.12 displays an extended analysis of the Norwegian results. The students are categorised in three scoring groups as before. Many responses in groups 1 and 2 are coded 70. Only among group 3 students is there a large percentage of correct responses. Looking at the curves for correct responses and the responses coded 70, we may characterise the distributions as complementary. It seems understanding that code 20/29 is correct is more or less the same as understanding that code 70 must be incorrect.

About 44 per cent of the Norwegian students scored one or two points. As induction is often considered as one of the most demanding content areas in school physics, this Norwegian result is quite encouraging. But as already mentioned, only the generally high-achieving students did succeed. Most students fail, *not* because of any complicated reasoning about induction, but because they do not see this task as a problem about induction at all! Seemingly, they are looking at the magnet and restrict the problem to the concept of direct magnetism in spite of aluminium being a non-magnetic material.

4.3   Water level with melting ice

> **G11**
>
> The water level in a small aquarium reaches up to a mark A. After a large ice cube is dropping into the water, the cube floats and the water level rises to a new mark B.
>
> What will happen to the water level as the ice melts? Explain your reasoning.

Archimedes' principle (or 'law') is usually stated as '*when a body is immersed in a fluid there is an upwards force which is equal to the weight of fluid displaced*'. This upward force is called the buoyant force and is a consequence of pressure increasing with depth. According to the principle, the water level in the aquarium remains the same because the ice displaces exactly the same volume of water as when it melts (namely the volume of water that has the same weight as the ice).

As with Newton's laws, Archimedes' principle is a fundamental concept in physics, and it is presented in science courses at different levels as if it were simple to understand. Also, young children in many countries are taught about floating and sinking and Archimedes' law. As we will show, even 'physics specialists' have great difficulties in applying the law or even in recognising when they should use this principle.

This item is of special interest because it touches on some environmental issues. Due to a possible higher global temperature in the future, ice can melt in the polar areas. But the consequences for the sea levels are very different whether the ice melts in the Arctic or in the Antarctic. Around the North Pole the ice is floating like the ice cube in the aquarium, and if the ice melts the sea level will remain the same. The consequences will be quite different if the ice on Greenland or in the Antarctic melts. Here the ice is lying on solid land and the sea level will arise if the ice melts. (Obviously there may be many other consequences if the global temperature rises).

Table 8.7. Item G11, Water level with melting ice: Coding guide

Code	Response
Correct Response	
20	Same level. Response refers to the fact that the volume (or mass) of the water displaced by the ice is equal to the volume (or mass) of the water produced when the ice is melted (Archimedes' principle)
29	Other acceptable responses
Partial response	
10/11	Same level. Incomplete, incorrect or no explanation
19	Other partially correct responses
Incorrect Response	
70	Rising level, with or without explanation
71	Sinking level. The water has smaller volume/greater density/ *'molecules are closer together'* than the ice OR the ice has greater volume/smaller density/ *'molecules are further apart'* than the water.
72/73/ 74	Sinking level. With other or without explanation
79	Other incorrect responses
Non-response	
90/99	

Table 8.7 is a revised coding scheme for this item and Table 8.13 shows the results from some selected countries. It appears that codes 70 and 71 are frequently used in most countries. Code 71 is particularly interesting. Responses in this category express the idea that the ice has greater volume, or that the molecules are further apart in ice than in water and therefore the water level will sink. In other words, these responses express the fact that ice has greater volume than water and that the volume will decrease when the ice melts. So far the argument is correct, but they do not see that the volume of the displaced water and the volume of the water produced when the ice melts are equal, and that consequently the water level will remain the same.

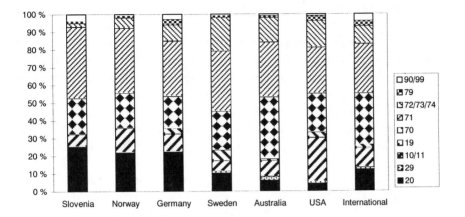

Figure 8.13. Item G11, Water level with melting ice: International and some national results

This is an example where many students express misconceptions or intuitive ideas. But it should be noticed that responses coded 71 involve partly correct thinking.

Also, for the two other physics items discussed here we have documented what are usually called misconceptions. However, we will argue that very often there is something correct in the 'incorrect' responses. It seems that there are some fragments or pieces of knowledge in the responses which in a way are correct. In our view, students do not have misconceptions that are constituting just naïve theories, but rather their ideas should be characterised as unstructured, fragmented knowledge. In this view the students are not constructing systematic and consistent theories, but different aspects or 'facets' of understanding are brought forward dependent on the actual context at hand (di Sessa 1993). The consequences for teaching should therefore be to build on this correct fragment of knowledge. Intuitive ideas do not need to be replaced so much as developed and refined.

5.   CONCLUSION

In this paper we have discussed some national and international results on selected free-response science items in TIMSS. The coding rubrics have functioned as an appropriate tool for describing student responses. We have demonstrated that we can obtain valuable insights into students' ways of thinking world-wide by analysing responses based on the coding rubrics. The coding scheme has proved to be sufficiently flexible to allow different ways of combining codes into categories, according to the special purpose and focus of the actual analysis.

All codes are strictly item-specific; they have been developed for a particular item with a particular phrasing. A revision of an item would demand revision of the codes as well. Likewise, each item is therefore analysed one by one. As long as the item contents are different, there is no reason to expect a particular combination of codes across items to be applied by the same students. On the other hand, the coding system is well-suited for exploring student responses to items that seek to assess similar concepts. In TIMSS, however, there are only a few examples of items similar in this respect.

Within a constructivist paradigm, students' alternative frameworks are often emphasised, thus implying that students develop consistent and somewhat stable conceptions. The consistency of the students' concepts is not easy to assess, but we will here draw attention to the fact that the ongoing TIMSS-Repeat study will provide an interesting opportunity to compare responses with almost identical, but still somewhat different, items.

The present discussion has focused on free-response items only. However, the multiple-choice items are also rich sources for diagnostic analyses.

The international reports published so far have focused attention on the attained curriculum and comparisons of scale scores and the influence on these scores of a range of background variables. It is now time to give more attention to the most important secondary analyses of the TIMSS database. The data are available for researchers in science and mathematics education, and it is up to this community to exploit this rich source of information for the benefit of improved science and mathematics instruction.

REFERENCES

Angell, C. and Kobberstad, T. (1993) *Coding Rubrics for Free-Response Items.* (Doc.Ref.: ICC800/NRC360). Paper prepared for the Third International Mathematics and Science Study (TIMSS)

Angell, C., Brekke, G., Gjørtz, T., Kjærnsli, M., Kobberstad, T., and Lie, S. (1994) *Experience with Coding Rubrics for Free-Response Items.* (Doc.Ref. ICC867). Paper prepared for the Third International Mathematics and Science Study (TIMSS)

Angell, C. (1995) *Codes for Population 3, Physics Specialists, Free Response Items.* TIMSS report no. 16, University of Oslo

Beaton, A. E., Martin, M. O., Mullis I. V. S., Gonzales, E. J., Smith, T. A., and Kelly, D. L. (1996) *Science achievement in the Middle School Years. IEA's Third International Mathematics and Science Study (TIMSS)* (Chestnut Hill, MA: Boston College)

di Sessa, A. A. (1993) Toward an Epistemology of Physics. *Cognition and Instruction*, 10 (2 and 3), pp. 105-225

Driver, R. og Easley, J. (1978) *Pupils and Paradigms: A Review Literature Related to Concept Development in Adolescent Science Students.* Studies in Science Education, 5, no. 61-83.

Ebison, M. G. (1993) *Newtonian in Mind but Aristotelian at Heart.* Science and Education 2, pp. 345-362.

Finegold, M. and Gorsky, P. (1991) *Students' concept of force as applied to related physical systems: A search for consistency.* International Journal of Science Education, 13, 1, pp. 97-113.

Kjaernsli, M., Kobberstad, T., and Lie, S. (1994) *Draft Free-Response Coding Rubrics-Populations 1 and 2* (Doc.Ref:: ICC864) Document prepared for the Third International Mathematics and Science Study (TIMSS)

Lie, S., Taylor, A., and Harmon, M. (1996) *Scoring Techniques and Criteria.* Chapter 7 in Martin, M. O. and Kelly, D. L. (eds) Third International Mathematics and Science Study, Technical Report, Volume 1: Design and Development (Chestnut Hill, MA: Boston College)

Martin, M. O., Mullis I. V. S., Beaton, A. E., Gonzales, E. J., Smith, T. A., and Kelly, D. L. (1997) *Science achievement in the Primary School Years. IEA's Third International Mathematics and Science Study (TIMSS)* (Chestnut Hill, MA: Boston College)

Mullis, I. V. S. and Smith, T. A. (1996) *Quality Control Steps for Free-Response Scoring.* Chapter 5 in Martin, M. O. and Mullis I. V. S. (eds) Third International Mathematics and Science Study: Quality Assurance in Data Collection (Chestnut Hill, MA: Boston College)

Mullis I. V. S., Martin, M. O., Beaton, A. E., Gonzales, E. J., Kelly, D. L., and Smith, T. A. (1998) *Mathematics and Science Achievement in the Final Year of Secondary School. IEA's Third International Mathematics and Science Study (TIMSS)* (Chestnut Hill, MA: Boston College)

Orpwood, G. and Garden, R. A. (1998) *Assessing Mathematics and Science Literacy.* TIMSS Monograph No. 4. (Vancouver: Pacific Educational Press)

Pfundt, H and Duit, R. (1994) Bibliography. *Students' Alternative Frameworks and Science Education.* 4th Edition. IPN at the University of Kiel, Germany

Robitaille, D. F., Schmidt, W. H., Raizen, S. A., McKnight, C. C., Britton, E. D., and Nicol, C. (1993) *Curriculum Frameworks for Mathematics and Science.* TIMSS Monograph No. 1 (Vancouver: Pacific Educational Press)

Robitaille, D. F. and Garden, R. A. (1996) *Research Questions and Study Design.* TIMSS Monograph No. 2 (Vancouver: Pacific Educational Press)

Sjøberg, S. and Lie, S. (1981) *Ideas about force and movement among Norwegian pupils and students.* Report 81-11. University of Oslo.

Third International Mathematics and Science Study (TIMSS) (1995a) *Coding Guide for Free-Response Items-Populations 1 and 2* (Doc.Ref.: ICC897/NRC433) (Chestnut Hill, MA: Boston College)

Third International Mathematics and Science Study (TIMSS) (1995b) *Coding Guide for Free-Response Items-Population 3* (Doc.Ref.: ICC913/NRC446) (Chestnut Hill, MA: Boston College)

Viennot, L. (1979) Spontaneous Reasoning in Elementary Dynamics. *European Journal of Science Education*, 1, 2, pp. 205-22

Wandersee, J. H., Mintzes J. J. and Novak, J. D. (1993) *Research on alternative conceptions in science.* in Gabel, D. (ed) Handbook of research on science teaching and learning (New York: Macmillan)

Chapter 9

Studying Achievement Gaps in Science within the National Context

Ruth Zuzovsky

School of Education, Tel Aviv University and Kibbutzim College of Education, Israel

1.   PERSPECTIVES

IEA studies have provided international comparative data on student achievement over a period of 35 years. The idea behind these studies was to look for the 'yield' of educational systems by testing the achievement of internationally comparable student samples.   .

The intensive work done by many research bodies all over the world (e.g., Organisation for Economic Co-operation and Development [OECD], the International Indicators of Educational Systems Project [IIES], the International Assessment of Educational Progress [IAEP]), has generated many publications and raised public interest in their findings.

Unfortunately, however, international comparisons of educational achievement began to resemble a horserace, something that provoked comments if not criticism. Critics pointed out methodological deficiencies, questioned interpretations given to results, and queried consequent policy decisions (e.g., Bracey, 1996, 1997; Baker, 1997; Stedman, 1997).

In addition to this criticism, it became evident over the years that within-country and within-state variance is much greater than between-state variance, and that the relative effect of variables explaining this within-state variance depends on the context in which they are grounded (Bracey, 1996). Thus, a gradual move away from decontextualised comparisons of achievements in favour of localised and contextual interpretation has occurred (Goldstein, 1996).

D. Shorrocks-Taylor and E.W. Jenkins (eds.), Learning from Others, 189–218.
© 2000 *Kluwer Academic Publishers. Printed in the Netherlands.*

In the light of this trend, this chapter presents local comparisons of achievement within the Israeli context and relates them to major policy changes of the last 30 years in the educational system in general, and in elementary science education specifically. Variables that were considered representative of prevailing policies in science education in the last 30 years were used as indicators of the functioning of this policy. The impact of these policy variables on science achievements was explored at three points in time, from the mid-1980s until the mid-1990s. This enabled an assessment of the implementation of the policies as well as the validity of their underlying assumptions. This processing of data obtained from large international studies exemplifies the move away from international comparisons toward localised and contextual comparisons of educational achievement.

1.1   Historical Description of Major Changes in the Educational Context in Israel

Education has always been held in high esteem in the Jewish tradition and in modern Israeli society. Israeli governments have consistently valued education for its own sake and as an instrument for social mobility and ethnic integration, something that is demonstrated by the country's relatively high expenditure on education.

The newly established country, which absorbed waves of Jewish immigrants from all over the world, recruited education to serve the aim of ethnic integration. The vision of statehood and the 'melting pot' rhetoric that dominated the political discourse in the early days of the state guided a policy that rejected the earlier liberal, religious and socialist Zionist ideological school streams in favour of only two separate Jewish sectors, i.e., the secular and the religious (with the exception of the ultra-Orthodox population who kept an independent education system) and one Arab sector. In 1953, the State took over the provision of a free, mandatory eight years of schooling, which used a uniform curriculum that emphasised modern Hebrew, homeland studies, and Biblical and Judaic studies. Toward the end of the 1950s, it became evident that the statist and uniformist policy did not attain its aim of social integration. Achievement gaps between Israeli-born and Ashkenazi-Jews from Europe and North America on the one hand, and Sephardi immigrants from North Africa and Asia on the other, remained, or even grew (Chen, Lewy, and Adler, 1978; Lewy and Chen, 1974). A differential compensatory policy which included the introduction of a remedial reading program, an extended school day and a secondary-level of vocational education, etc., was implemented for a period of ten years from 1958 to 1968 (Schmida, 1987). The failure of this strategy led, in 1968, to another major educational reform: a change in the structure of the educational system from an 8 + 4 year system to a 6 + 3 + 3 year system, the extension of free mandatory schooling from kindergarten to the age of 18, and the abolition of standardised selective tests at the end of the 8th grade, all aimed to pave the way for secondary education to all. The establishment of the middle/junior high school contrived, by bussing students from different areas, to implement the policy of imposed social integration.

This policy of integration was met with objections from different interest groups, especially from parents in wealthy communities. Even today it has not been fully implemented. Nevertheless, the junior high-school, with its heterogeneous population, became the locus of experimentation for new, centrally developed curricula and pedagogical strategies that aimed to overcome individual differences (by means of mastery learning models, differentiated programs for slow children and for talented children, etc.).

The two aims of advancing the disadvantaged (no more talk about 'ethnic gaps') and of the fostering of achievement, although in mutual conflict, constituted the flagship of the Israeli educational system in the 1960s and 1970s. Coinciding with an international trend of curricular reforms of an experimental nature which built upon notions concerning the structure of disciplinary knowledge, this period can be viewed as the curricular reform period in the fledgling Israeli education system.

Yet another trend, resulting largely from technological as well as social and economic changes in Israel and the western world as a whole, during the 1980s and 1990s, marked a new period in education. Advances in telecommunications, dissemination of knowledge, the move to hi-tech industries, the need for flexible organisation and decentralised decision making, and the erosion of ideological as well as scientific certainties, have all had their impact on education. These trends, together with the election of a right of centre government which favoured a free market policy, privatisation and decentralisation, led to more autonomous local school districts and encouraged individual schools to assume administrative, economic and curricular responsibilities.

The centralised educational system, however, still maintains some overarching priorities. Two examples of a centrally led policy are related to the achievement comparisons presented in this chapter. The reform in teacher education can be traced back to the structural reform of schooling. One is the reform in teacher education, a consequence of the 1968 structural reform, and the second is the recently implemented reform in science education. Once the junior high school was established, a call for the upgrading of teacher education was made by both the government and the teacher unions. In 1979 a governmental committee (the Etzioni Committee) set in motion the development, from 1981, of a model for the academization of all teacher education institutes (Dan, 1983). By 1994/5, half the teacher training colleges in Israel were offering studies toward academic degrees.

The other centrally led policy is the 5-year plan, 'Tomorrow 98', that aimed to foster science and technology education. This chapter examines in detail its implementation and impact during its early years.

1.2   Reforms in Science Education in Israel

If we now narrow our focus to science education, the first thing to be observed is that reforms in science education in Israel always take the shape of curricular reform. A historical review of curricular trends in Israel (Ben Peretz, 1986) shows that there have been three generations of curricular activity since the foundation of the state: an early period (1948 to 1963) characterised by centralised, expert development of traditional textbooks; an intermediate period – the 'scientific' period – in the 1960s and 1970s which was inspired by the work of the Chicago school and Schwab's ideas about the structure of the discipline. This period too was marked by experimental curricular planning, now by teams of collaborators rather than by one expert. The third period, stretching from the late '70s onwards, could be called the 'social humanistic' period, which emphasises the societal relevance of learning and the personal nature of knowledge construction. The curricula so far developed during this period have aimed at learners' active involvement in societal and environmental issues. This approach was in line with the concurrent, earlier mentioned, local development of school-based curricula. Another feature that distinguished the curricula developed in the different periods was the students they served. In the first period, students were viewed as a homogenous body studying from uniform learning materials. The second period was more sensitive to individual differences, and this resulted in curricula aimed toward a heterogeneous population. At present, curricula aim, again, at the broad student population, distinguishing between literacy demands for all and academic demands for selected populations.

Three similar distinct periods can also be perceived in the evolution of the Israeli elementary science curriculum. The first of these is the pre-state curriculum that focused on natural science and homeland studies. Traditional, single author textbooks represented that period. The second period, after the establishment of the state, and after the institution of the official Centre for Curriculum Development, was dominated by the scientific-inquiry approach in curricular development. In the 1970s, the MATAL curriculum – science for elementary schools – was developed and fully implemented in all national elementary schools. This 'new' curriculum centred on disciplinary structures of knowledge and emphasised the acquisition of scientific processes in addition to the acquisition of scientific content knowledge. The new program advocated exploratory, investigative teaching methods rather than frontal expository ones. Its main objective was to cultivate a generation of future scientists. During the years of its implementation, a less élitist, integrated Science Technology and Society curriculum, in line with an international trend toward scientific literacy for all, emerged.

From the 1980s on, a new curriculum – MABAT (Hebrew acronym for Science in a Technological Society) – replaced the previous MATAL curriculum, becoming the locally dominant elementary science curriculum. One of the main aims of this curriculum, as opposed to the previous one, is to 'educate informed citizens to utilise science and technology for improving their own lives as well as for dealing responsibly with science related social issues' (Chen and Novick, 1986). This objective fits in with what is currently often defined as scientific and technological literacy. In contrast with the MATAL curriculum which was very process-oriented and experimental in nature, this curriculum is less laboratory and science oriented and focuses more on the acquisition of thinking skills, such as problem solving and decision making, and in the social consequences of science and technology.

1.3   Science Education Policy in Israel – at the end of the twentieth century

In spite of the very intensive and innovative science education activity which for years has characterised science education in Israel and in spite of constant curriculum reform, the educational yield of this activity has not been impressive. In 1984 Israel ranked eight among eleven countries that participated in the 2nd International Science Study of the IEA at the elementary level (Keeves, 1992). This fact, and the low enrolment levels in upper elementary advanced physics, chemistry and even biology, together with the forecast of a dramatic change in the work force and in the skills needed to cope with the technological advances of the information age, led the Minister of Education in 1990 to appoint a committee headed by the President of the Weizmann Institute of Science, Professor Harari, whose task it was to 'offer suggestions for educational programs, special projects, pedagogical and organisational improvement and any other initiative to advance science and technology education toward the 21st century' (the appointment letter to the members of the committee, 1990).

In 1993, a five-year plan, 'Tomorrow 98', which referred to both mathematics and science studies at all grade levels, was launched. One recommendation included in the plan dealt specifically with science at the elementary level; here the committee recommended 'improved, extended and in-depth study of science; minimum weekly hours per student; revision of curricular materials; introducing science and technology labs; a comprehensive system of teacher preparation in teacher colleges, and through in-service training; introduction of science and technology co-ordinators in schools, and establishing national and regional support centres for teachers implementing the program ('Tomorrow 98' Report, 1992, p. 28). These recommendations were implemented nation-wide from 1993 onwards.

1.4   Achievements and Policies

So far, this chapter has provided a historical look at the context against which the science achievement of elementary students at three points in time can be interpreted. In the main part of the chapter we present achievement comparisons and relate them to the changes in policies described above and to their level of implementation. Conclusions drawn from this analysis enable an assessment to be made of the underlying assumptions and effectiveness of these policies. This analysis is needed especially in light of the unsatisfactory recent achievement of 4th and 8th grade Israeli students in the 1995 Third International Mathematics and Science Study (Martin, Mullis, Beaton, Gonzalez, Smith and Kelly, 1997; Zuzovsky, 1997).

1.5   Data Source

The data to which this chapter refers were obtained from three nation-wide surveys. The first one was carried out in Israel in 1984 as part of the Second IEA Study on Science Achievement (SISS) (Zuzovsky et al., 1987). The second batch of data was obtained in 1992 through a follow-up study conducted in the same schools that had been involved in the 1984 study, using short versions of 1984 questionnaires (Zuzovsky, 1993). The third survey was done more recently (1995) as part of the Third International Mathematics and Science Study (TIMSS). This time, new instruments with comparable items, which were matched and recoded, were used again in the same schools that had already participated in the past.

The school sample used for the 1984 analysis comprised 86 schools. It was a stratified random sample from the population of 1,110 elementary Hebrew schools. One fifth-grade class was randomly chosen from each school and all the students in that class, their science teacher and the principal participated in the study. In 1984, complete data were obtained from 69 schools, 69 classes and 1,528 pupils. A science test with 45 core items was used as the dependent variable and a reading comprehension test score was used as a proxy of pupil's ability.

In 1992, the same sample was used. Two 5th grade classes (instead of just one) were sampled randomly from each school. This was done in order to represent more adequately class level variables which had been confounded with school level variables in the previous study. Complete data were obtained from 75 schools, 140 classes and 3,088 pupils. Of these 75 schools, 69 had been previously sampled in 1984, so repeat measures in these schools were obtained. A score on a science test consisting of 37 core items which had appeared among the 45 core items in 1984, and an additional 28 new items, made up the dependent variable. A score on the same reading comprehension test administered in 1984 was used again as the pupil ability surrogate.

In 1995, 87 schools from an original sample of 100 schools participated. Forty-seven of them were replaced with matched schools that had already participated in the two previous studies. These schools were thus revisited for the third time. One 4th grade class was randomly chosen from each school and the students of that class, their science teacher, and the principal of that school took part in the study. The number of students participating in this study was 2,351, the number of participating science teachers was 78. Only 50 out of 78 participating schools principals responded, which left the school level data file incomplete.

Nine forms of tests were composed from 199 test items of which 97 were science test items and the rest were mathematics items. In some forms there were more science test items than in others, but the forms were equally distributed in each class, so that the class mean of science achievement score can be used as a comparable achievement measure at class level. This time the mathematics score, not a reading comprehension score, was used as the pupil ability surrogate.

1.6   Method

The collection of data at three points in time enabled the preparation of a longitudinal file of input and process variables related to students, teachers, instruction, learning and schools. All variables used at the three points in time were recoded and converted, if necessary, to have the same meaning and identical scales (categories and direction) over the years. The variables describing students' characteristics were derived from students' questionnaires, and those describing class- or school-level variables mainly from teachers' or principals' questionnaires. However, some variables representing class level characteristics, such as instruction, are student level variables aggregated to become class level variables. Frequency distribution of these variables across the three points in time reveals changes that occurred over time.

Changes in the pattern (size and direction) of the relationship between these variables and science achievement was determined by means of Pearson correlation coefficient. A breakdown of mean achievement percentage scores, as well as standardised T-scores of sub-populations defined according to the categories of the policy variables, made it possible to distinguish both linear and non-linear trends of change in the three sets of data. For instance, the mean achievement score of students from small, medium or large families was found to have a similar pattern of change over time: a decrease in the achievement of students coming from large families. However this achievement gap was found to be narrowing over the years – a finding that needs to be further explored. Table 9.1 illustrates this.

Table 9.1.

	1984		1992		1995	
Size of Family	Raw per cent score and (SD)	T-score	Raw per cent score and (SD)	T-score	Raw per cent score and (SD)	T-score
Small family	64 (17)	53	54 (16)	51	65 (18)	51
Medium family	61 (17)	51	53 (16)	51	65 (17)	51
Large family	53 (17)	46	48 (15)	47	59 (17)	48

This chapter presents changes over time that occurred in a selected set of input and process variables. The selection of variables was guided by their role as alterable policy variables, theoretical considerations and previous empirical findings on their effect on science achievement. The variables are classified into the following two main groups:

1. *Input Variables* i.e. relating to pupil's background, teacher's background and school's structural and material conditions. These variables are considered covariates of educational achievement.
2. *School Level and Class Level Process Variables* that represent preferable processes according to the prevailing science education policies.

Data concerning frequency of the input and process variables, correlation with science achievement scores, and breakdown of mean achievement scores of sub-populations defined according to categories of the selected variables shed light on changes that occurred from 1984 to 1995 in the effectiveness of the schools' science education.

2.   FINDINGS: CHANGES IN THE LEVEL AND EFFECT OF INPUT AND PROCESS VARIABLES RELEVANT TO SCIENCE ACHIEVEMENT IN ISRAEL FROM 1984 – 1995

To interpret the meaning of these changes, we calculated the size and direction of the correlation coefficient of each variable with science achievement, and the mean achievement and the standardised increment above or below the mean of the sub-population defined according to the variable categories.

2.1   Findings Related to Input Variables

2.1.1   Pupils' background characteristics

Several variables describing students' socio-economic, demographic, as well as attitudinal characteristics were used to trace changes in student body composition. These variables were known from previous studies (many of which are cited in Shavelson, McDonnell and Oakes (1989)) to have an effect on learning achievement:
– Family size (1 = small [3-4 people], 2 = medium [5-6 people], 3 = large [more than 6 people])
– Number of books in the student's home (1 = few, 2 = up to 100, 3 = 100 to 200, 4 = more than 200)
– Students born in Israel (0 = not born in Israel, 1 = born in Israel)
– Number of generations student's family has lived in Israel (0 = not born in Israel, 1 = first generation, 2 = second generation)
– Positive self-concept as science learner (1 = low, 2 = medium, 3 = high)
– Positive attitude toward the study of science (1 = do not like, 2 = like, 3 = like a lot)

Usually one extreme category of these variables served as an indicator of the variable. The percentage over time of students falling in that extreme category enabled us to portray changes that occurred in these indicators over the years (see Table 9.2).

Correlation coefficients of student background variables with science achievement and a breakdown of mean achievement of subpopulation defined according to subcategories of each variable and an increment above or below standardised mean enabled us to reveal the relationship between these variables and science achievement. For details see Appendix 1 to this chapter.

Table 9.2. Changes (1984-1995) in Selected Indicators of Students' Background

Indicators of Students' Background Variables (Variable category used in brackets)	Per cent of Students		
	1984 n = 2216	1992 n = 3525	1995 n = 1908
Students from large families (3)	23	18	14
Students from homes with many books (4)	46	24	21
Israeli-born students (1)	92	90	88
Second generation Israeli students (2)	43	60	66
Students with high positive self-concept as science learners (3)	12	18	51
Students liking science a lot (3)	25	30	43

The results indicate a decline over time in the percentage of students from large families; an increase in the percentage of second generation Israeli students; an increase in the percentage of students with high positive self-concept as science learners, and an increase in the percentage of students who claim that they like science. All these tendencies have a positive effect on science achievement. On the other hand, there is a decline in the proportion of students coming from homes with many books and a decline in the percentage of native Israelis. These tendencies have a negative effect on science achievement. The opposing demographic tendencies seem to compensate for each other. The growth in the number of immigrant students, which has a negative effect on achievement, is compensated by the increased number of second generation Israeli students and by the decline in proportion of students from low socio-economic homes (large families), both having a positive effect on achievement. These balanced changes in students' variables and the general pattern of a decrease in the achievement gap attributed to them lead us to conclude that student body composition, although slightly changed, should not be considered as a factor that explains the drop in student outcomes.

2.1.2   Teacher background variables

Teachers' education level, their academic and pedagogical preparation and their years of teaching experience are regarded as indicators of their proficiency. In the case of Israel, the move from a two-year preparation program at a teacher seminar to a three-year preparation program, and then to a four-year academic program, mark an improvement in teachers' education. Combined academic and pedagogical preparation done either simultaneously as in teacher colleges, or sequentially as in the universities, is

the preferred mode of preparation.   Several indicators were chosen to identify improved teacher education, as follows:
- Percentage of teachers with two years of preparation.
- Percentage of teachers with three years of preparation.
- Percentage of teachers who had been in academic programs (received academic degrees).
- Percentage of teachers with both academic and pedagogical preparation.
- Percentage of teachers with more than 20 years of teaching experience.
- Teachers' professional behaviour as indicated from investing time in activities beyond classroom teaching such as planning lessons and in time allocated for professional development activities: reading professional literature, in-service training, updating, etc.   Two levels of time investment in these things were determined: 1 = less than three hours per week, 2 = more than three hours per week, a sign of professionalism.

Table 9.3. Correlation Coefficients of Teachers' Background Variables with Science Achievement and Achievement Means of Sub-populations Defined According to Categories of Teachers' Background Variables and Increment Above or Below Standardised Mean

	1984 n = .78			1992 n = .77			1995 n = .75		
Indicators of Teachers' Background Variables	Corr. Coeff.	Mean (SD)	Incre.	Corr. Coeff.	Mean (SD)	Incre.	Corr. Coeff.	Mean (SD)	Incre.
Two years seminar	-.16	60 (7)	0	.04	51 (7)	0	.01	63 (7)	0
0 = other; 1 = two years		57 (11)	-1		52 (9)	0		63 (2)	0
Three years seminar	.14	58 (9)	-1	-.15	53 (7)	+1	-.09	63 (6)	0
(0 = other; 1 = three years		61 (8)	+1		50 (8)	-1		62 (7)	0
Academic degree	-.03			.10			.07		
0 = without academic degree;		59 (10)	0		51 (9)	0		62 (7)	0
1 = with academic degree)		58 (2)	-1		54 (7)	+1		63 (7)	0
Combined	-.02			.30*			.21		
academic/pedagogical		59 (9)	0		50 (7)	-1		62 (7)	-1
preparation									
0 = other; 1 = combined		58 (3)	-1		57 (6)	+3		65 (6)	+1
preparation									
Experienced teachers	-.02	60 (8)	-1	.03	51 (8)	0	.25*	62 (7)	0
(1 = up to 5 years; -4 more		59 (11)	0		51 (7)	-1		59 (7)	-2
than 20 years)									
		61 (8)	+1		52 (8)	0		63 (6)	0
		57 (9)	-2		52 (7)	0		67 (4)	+2
Teachers allocating time for	.07			.15			.11		
extra teaching – planning		58 (7)	-1		48 (8)	-2		61 (6)	-1
1 = less than 3 hours; 2 =		60 (9)	0		52 (8)	0		63 (7)	0
more than 3 hours									
Teachers allocating time for	.27*			.04			.10		
professional development		57 (9)	-2		52 (6)	0		60 (7)	-1
1 = less than 3 hours; 2 =		62 99)	+1		53 (10)	0		64 (6)	+1
more than 3 hours									

Corr. Coeff. = Correlation coefficient

Incre. = Increment T-score – Points above or below standardised mean

Substantial changes in the characteristics of elementary science teachers have occurred between 1984 and 1995: there has been a decline in the numbers of teachers with only two years of preparation, a doubling of the number of teachers with three years of preparation, and a dramatic rise in the percentage of teachers with an academic degree. By 1995 one third of the total population was in possession of a university degree. A slight increase in the number of experienced teachers was also observed. However this was the effect of ageing, since we observed a constant ratio of 50 per cent, since 1984, of practising teachers with more than ten years of experience. (For details see Appendix 2 of this chapter.)

These changes might explain the direction of the correlation coefficients of the teacher variables with science achievement (Table 9.3). This Table includes a breakdown of class means of science achievement of sub-populations defined by the categories of teacher variables. In 1984, when there were no exclusively academically schooled teachers in Israel, it was the three-year level of preparation which was positively associated with achievement. In 1992 and 1995, with the academization of the teaching force on the increase, the academic and the combined academic and pedagogical preparation appear to be highly correlated with learning outcomes.

To sum up, changes that occurred in the teacher background variables during the last ten years point towards an improvement of this professional force. Combined academic and pedagogic preparation and years of experience are indeed proven indicators of the proficiency of teachers.

Regarding teachers' professional activities beyond teaching in the classroom, a dramatic increase in the time which teachers allocate to in-service training has occurred. This increase reflects a policy implemented since 1994 of fostering school-based and school-initiated in-service courses, by offering all teaching personnel at least two hours of in-service education weekly. Participation in in-service courses was recognised and credited. An increase in the time teachers allocate to learning in an in-service framework marks an upgrading of teachers' professionalism. Almost all teachers report allocating more than three hours per week for in-service courses – a remarkable percentage even when compared with other countries.

2.1.3   School input variables – structural or contextual

We identified the following school variables: 6-year school (1), or a school with the old 8-year structure (2); size of school: small – (1) with one or two parallel grade classes, medium – (2) with three to four parallel grade classes and large – (3) with more than four parallel grade classes; and type of community in which the school is nested: rural – (1), suburban – (2), inner city – (3).

The findings show that by 1995, 77 per cent of the schools were 6-year schools, i.e., almost 20 years after the initiation of structural reform in the educational system, it had not been fully implemented. The findings also show a small and inconsistent advantage of 6-year schools with respect to science achievement. In 1995 there were fewer small schools and fewer inner-city schools than in the previous years. (For details see Appendices 3 and 4 of this Chapter.) This may have been due to a different sampling procedure used in 1995. There is no clear-cut evidence regarding the impact of these structural features of schools and the achievement gaps attributed to them vary but, on the whole, they are not large.

2.1.4   School input variables – physical conditions

Two indicators of physical conditions over years were chosen for our comparisons. They related to recently highly prioritised policies in science education world-wide and a main recommendation of the 1993 Harari Report in Israel which entailed supplying science and technology laboratories with materials and equipment and introducing computers into schools. Information on the conditions regarding laboratory equipment and computers was obtained indirectly in the Israeli surveys. In 1995, principals were asked to what extent they considered their schools' instructional capacity was affected by the inadequacy of science laboratory equipment and materials. In 1984 and 1992 teachers were asked to what extent they found implementation of the science curricula to be constrained due to inadequate working conditions in the lab. These responses to material constraints were used as indicators of the physical conditions of science teaching.

The research reveals a drop between 1984 and 1995 in the percentage of schools in which the principals, or teachers, view science laboratories as inadequate. (For details see Appendix 5 to this chapter.) This variable ceases to be negatively correlated with science achievement in 1995, which might reflect the reduced role of practical work according to the new STS curriculum. Another school condition indicator related to the Harari report is the number of computers in a school, but information about this was only obtained in the 1995 survey, when the average number of computers for instructional purposes in school was 17. Twenty per cent of schools did not have any computer for instructional purposes.

Data obtained from Ministry of Education publications (Israel Believes in Education, 1996) support the above findings, reporting a ratio of one computer per 29 students in 1992 and one computer per 19 students in school in 1995. However, the number of computers in school does not correlate significantly with science achievement

2.2   Findings Regarding Process Variables Related to the 'Tomorrow 98' Policy

A set of variables describing processes at the class and school levels, which are in line with the recommendations of the Harari Committee, were selected to serve as indicators representing this policy. The school level variables are as follows.

–   Level of 'new' science curriculum implementation: (1 = not at all or partial; 2 = full implementation). In 1995 this variable had a different meaning which reflected decentralisation trends: Principals were asked whether the schools have their own stated science syllabus in addition to the national one (1 = yes; 2 = no).

–   Level of integration of science studies with other curricular subjects: Data on this variable were obtained only in 1992 and 1995 (0 = not integrated; 1 = integrated).

–   Provision of within-school enrichment programs: In 1984 principals reported on the existence of science clubs (1 = yes; 2 = no), in 1992 on special projects in school (1 = no; 2 = yes) and in 1995, students responded whether they participated in science or mathematics club (1 = yes; 2 = no).

–   Out-of-school science activities: In 1984 – occurrence of trips: (1 = yes, 2 = no). In 1992 – (1 = no; 2 = yes). There were no items referring to this indicator in 1995.

–   The extent to which computers were actually incorporated and used in learning and teaching: In 1984 teachers were asked this question; in 1992 the respondents were principals and in 1995 students responded. All responses were recoded to 1 = often; 2 = rarely.

–   School discipline and atmosphere – the extent principals or school personnel have to deal with discipline problems in school: In 1984 and 1992 principals responded to one general question on this issue, while in 1995 principals responded to a battery of related questions. One question of this battery on the frequency of classroom disturbance served as an indicator. Responses to this question were given on a scale ranging from 1 = seldom, 2 = sometimes (once in a while) 3 = often (once a week, a day).

The findings are that only four process indicators in line with the recommendations of the Harari Report show growth patterns: level of implementation of the official science curriculum or developing schools' own science syllabus: provision of enrichment programs: the use of computers in classrooms, and an increase in the amount of time teachers devote to professional development (in-service courses). These variables are positively correlated with science achievement. (See Appendices 6 and 7 of this chapter.)

The impact of a trend toward more school autonomy and responsibility is evident. In 1995, one third of the schools developed and implemented their own science curricula. This variable indicates a general policy supporting school-based curriculum development.

Contrary to the recommendation of the Harari Report, science continues to be taught as a separate subject and is not integrated with other subjects. However, the difference in achievement between groups which study science as an integrated subject and as a separate subject is minor, so that this recommendation does not seem to be empirically grounded and appears in need of further evaluation.

2.3   Instructional Practices

Three main types of instruction that aim to foster student learning are practised in science lessons: whole class teacher-led learning, small group learning, and individualised learning. 'Tomorrow 98' aimed to improve students' understanding, motivation to study science and interest in studying it by an increased use of student-centred teaching approaches, individual learning and an enhanced use of computers in learning. Data on the frequency of these modes of learning and instruction were obtained from students and teachers. Student responses were related to specific classroom activities: copy notes from the board, have a quiz or test, etc. Teacher responses, on the other hand, were related to general working habits. Student responses were given on a 3-point scale (1 = often; 2 = sometimes, and 3 = never), while teacher responses were on a 4-point scale (1 = never; 2 = sometimes; 3 = mostly, and 4 = every lesson). These ranges can be used to indicate the source of the variables. Only one category on these scales (the most frequent) served as an indicator variable.

The following variables were used as indicators of whole class teacher-led instruction.
–   Copying notes from the board (1-3)
–   Teacher checks homework (1-3)
–   Teacher demonstrates an experiment (1-3)
–   Teacher answers questions raised by pupils (1-3)
–   Teacher lectures (1-3)
–   Working as a whole class with teacher leading (1-4)

Another set of variables was used as indicators of individualised learning and instruction:
– Students do experiments on their own (1-3)
– Students work individually without assistance (1-4)

Variables used as indicators of small group learning and instruction are as follows:
– Working pairs or small groups (1-3)
– Working in pairs without teacher assistance (1-4)

Appendix 8 presents the percentage of classes/schools, at three points in time, where one of the three types of instruction is frequent. When the information regarding instruction was derived from teachers' questionnaires, it yielded the percentage of classes/schools where the type of instruction occurs every lesson (category 4). In the case of information derived from students' questionnaires, the variables were dichotomised. This gave the percentage of schools with means on the instructional scale falling below 1.7 (i.e., representing a high occurrence of the instructional mode). The relationship between the indicator variable and science achievement, i.e., correlation coefficient and breakdown of achievement (class means) according to the variable categories are presented in Appendix 9.

The findings showed an increase between 1984 and 1995 in the percentage of classes in which whole class, teacher-led instruction occurs. The percentage of schools exhibiting a high frequency of students copying from the board and routine checks of homework by the teacher increased. Because experimentation became less frequent in STS science-oriented lessons, more teachers restricted themselves to demonstrations. The relationship between these variables and science achievement is not consistent. Many of the whole class instruction indicators are positively and consistently associated with science achievement. However, some manifestations of teacher-led instruction, e.g., copying from the board, teachers showing how to solve problems, although more frequent in 1995, seem to be negatively correlated with achievement.

Typical practices in science lessons are self-study, individual work on tasks and doing experiments. In general, there is a decrease in teachers' reports on individualised instruction and self-experimentation. However, the relationship of this type of instruction – which is that recommended in 'Tomorrow 98' and other national policies – with science achievement, is positive.

The frequency of group work which used to be very popular according to the inquiry-oriented curriculum in the 1980s, is declining. However, the positive association with science achievement remains.

When we sum up the pattern of change in instructional modes in science classrooms over the period under study, several trends can be perceived viz.:

- An increase in whole class teacher-led types of instruction, which is usually positively associated with science achievement.
- A decline in individualised type of instruction and self-experimentation of students which are still positively associated with science achievement.
- A decline in group work, a type of learning which seems in 1995 to be less associated with science achievement.
- Two thirds of teachers report using all modes of instruction in some lessons. This leads us to conclude that teachers tend to diversify their style of instruction.

3.   CONCLUSIONS

This chapter has provided longitudinal comparisons of science achievement at the national level in the context of several reforms that occurred in Israel since the late 1960s. This includes general reforms and reforms related specifically to science education. Among the general reforms there was a structural reform that changed the system from an 8 + 4 structure to 6 + 3 + 3 year structure, and academization in teacher education, both pre- and in-service. The specific reforms in science education touched curriculum, pedagogy and the infrastructure of science teaching in schools. The present comparisons enable an evaluation to be made of the level of implementation of all these reforms and the validity of some of the assumptions that underlie them.

The structural reform envisioned the phasing out of the 8-year elementary school and the rise of the 6-year school. Was this reform completed? Did it have any effect on student outcomes? As can be seen from the data obtained, this reform was not fully achieved. In 1995 only 77 per cent of elementary schools were 6-year schools and the level of attainment of students enrolled in these schools does not significantly differ from those of students in 8-year schools. Continuous objection to the social integration policy behind this reform and unconvincing findings regarding the social and learning benefits of this policy in the junior high level (Chen, Lewy, Adler, 1978), led to the idea of attaching the junior high-school to the secondary school, thus creating a new structure of 6 + 6 grades, a process that had already started.

Along with the structural reform in the educational system came also curriculum reform and reform in teacher education. On repeated recommendation (the Peled Commission, 1976, the Etzioni Commission 1979), the academization of teacher education institutes started. In 1981 the Commission of Higher Education provided guidelines (Dan, 1983) related to the structure, content, curriculum, admission requirements and upgrading faculty profiles in teacher colleges which were left to interpret and apply these guidelines. A tedious process of accreditation followed this effort. In 1994/5, over half of the student-teacher population in 15 out of 34 teacher training colleges in Israel were enrolled in studies for an academic degree (B.Ed.). Parallel to the academization of pre-service education, a policy that encouraged working teachers without a university degree to continue studies toward a B.Ed., was initiated and the number of practising teachers participating in associated B.Ed. programs grew from 744 in 1989 to 7400 in 1995. To what extent had the academization process an effect on student achievements? This question cannot be fully answered from the data gathered in this study. However, a small advantage in science achievement of students who were taught by teachers with an academic degree can be observed. This advantage increases when students are taught by teachers with combined academic and pedagogical training. It seems, then, that this part of the reform (academization of the teaching force) is effective with regard to student outcomes.

Reform in science education in Israel followed the world-wide curricular reforms during the 1960s. The inquiry-oriented science curriculum of the 1970s gave way to an interdisciplinary curriculum that integrated science, technology and social studies (STS). This curriculum was fully implemented by the early 1990s, so that the 1984 achievement data reflect the impact of the 'old' disciplinary curriculum, while the effect of the 'new' STS curriculum appears in the 1992 and 1995 data. The decline in Israel's position in international studies and on common core items that appeared both in the 1984 and 1995 tests does not support the claimed advantage of the STS curriculum over the previous discipline-based programme.

The Harari report published in the early 1990s and the subsequent 'Tomorrow 98' Program launched in 1993, shifted the emphasis from curricular development and its implementation to teachers and instruction at the elementary level. Most of the money was redirected from curricular development to in-service training. Teachers were trained to teach the new, integrated curriculum, to use student-centred teaching approaches in their instruction, to foster students' individual learning, and to use computers in their instruction. Another, related, policy encouraged teachers to participate in central as well as school-based in-service courses. In 1995, 72,000 teachers out of 85,000 were enrolled in a variety of in-service courses. It is too early to assess the impact on student learning of this policy of in-service programs, which was only implemented in 1994. However, according to an evaluation report from the Ministry (Evaluation Unit, 1996) there is an increase in the percentage of both teachers (from 20 to 45 per cent) and principals (from 45 to 62 per cent) who think the program contributed to improving the school climate and an increase in the percentage of teachers (from 15 to 25 per cent) who think that the in-service program contributed to improved student outcomes. Our data support these findings.

Equipping schools with computers and establishing science laboratories was another aspect of the Tomorrow 98 reform in science education. A ratio of one computer per ten students in the primary and junior high school was set as the ultimate goal. The program also aimed to establish and operate fully equipped science laboratories with a view toward maintaining continuous connection between theoretical studies and experimental activity, so as to integrate technology in science learning. The objective was that at the end of the program every student would participate in two to three hours of laboratory activities each week. This seems not to have been achieved.

The data gathered in this study make it possible to examine critically the goals of Tomorrow 98, the level of their implementation and their impact. Tables 9.4 (a) (b) (c) and (d) summarise the implementation of the main goals of Tomorrow 98 (relating to schools, science teachers, type of instruction and pupils) and their association with science achievement.

Table 9.4(a)

School Level		
Recommendations and Policy Guidelines	Level of Implementation	Impact/ Effect
Establishing infrastructure of laboratories and equipment	In almost half of school principals report on inadequate conditions regarding labs and equipment	No clear effect
Increase in science teaching hours	Though allocation of teaching hours increased, they were not directed to science teaching	Positive effect
Computerising schools	Average number of computers in school is 17. In 18% of schools, there are more than 30 computers	No clear effect
Actual use of computers in classroom in classroom instruction	Only 11% of students report on frequent use of computers in their classroom instruction	No clear effect
Developing and implementing school-based science curriculum	In more than a third of schools	Positive effect
Integrating science with other school subjects	In almost all schools, science is taught as a separate subject	No clear effect
Providing remedial instruction in science	In almost one quarter of schools	Positive effect
Providing extra-curricular science enrichment activities	In almost half of schools	Positive effect
Disciplined atmosphere	More than one quarter of schools has to frequently deal with classroom disturbances	Large negative effect

Table 9.4(b)

The Science Teacher		
Recommendations and Policy Guidelines	Level of Implementation	Impact/ Effect
Academization of teacher education	32% of elementary science teachers hold an academic degree; 25% also have pedagogical training; very low percentage in comparison with many other countries	Positive effect
Promoting in-service science training	70% of elementary teachers devote more than three hours weekly to in-service training; high percentage compared with other countries	Small positive effect
Years of experience	About 60% of teachers have more than 10 years of teaching experience. Average experience years of Israeli science teacher is less than elsewhere.	Positive effect
Investing extra classroom time	The Israeli science teacher devotes additional 16 hours per week for extra-instructional activities especially lesson planning; this is much more than elsewhere.	Positive effect
Meeting with other teachers	About half of the teachers meet regularly, weekly for discussion; smaller percentage than elsewhere.	Positive effect
Familiarity with the official national syllabus	Half of the teachers know the national science syllabus very well; much more than in other countries.	Small positive effect

Table 9.4(c)

Instruction Level		
Recommendations and Policy Guidelines	Level of Implementation	Impact/ Effect
Teacher led frontal instruction	In 42% of the classrooms (in part or in almost all of the lessons)	Small positive effect
Individualised instruction	Occurs in about 87% of the classrooms (in most lessons)	Small positive effect
Groupwork without teacher assistance	Occurs in 27% of the classrooms	Small positive effect
Alternative ways of instruction	Frequent work on projects or on everyday problem solving in only a few classrooms. Three modes of instruction are used with individuals rather than with the whole class	No clear effect
Laboratory work	Teacher demonstrates experiments in 1/3 of the classrooms	Small positive effect
	In only 6% of classrooms, students do experiments in most lessons	Positive effect

Table 9.4(d)

Pupil Level Variables		
Recommendations and Policy Guidelines	Level of Implementation	Impact/ Effect
Demographic variables	Unchanged over the last ten years. Most pupils come from medium sized families. In half of the homes, there are more than 100 books. Two thirds of the pupils are second generation Israelis.	Family size is positively associated with learning outcomes and so is the number of books at home.
Attitudes	Half of the students perceive themselves as doing well in science and 43% of them really enjoy studying science	Positive effect
Locus of control	One third of the pupils do not agree and more than a third definitely do not agree that luck plays a role in succeeding in science studies.	Positive effect

Some conclusions can be drawn, if we return to the use of contextualised achievement comparisons and infer from these comparisons the effectiveness of the policies that guided these changes.

At the student level some demographic changes appeared, most of which were related to the increased number of students from newcomer families. This trend negatively affects achievement. However this is compensated by other indicators showing that students' interest in science and their self-concept as science learners increased. These indicators are positively associated with achievement and high levels of professional behaviour. In spite of better teacher preparation, teacher-led instruction continues to be the most common type of instruction, while self-experimentation, which is positively associated with achievement, is on the decline. The popularity of group learning has dropped and was found to be negatively associated with achievement (see Table 9.5).

At the school level, some characteristics that are positively associated with achievement appear to be the more frequent use of computers, enrichment activities and the supply of appropriate laboratories and materials. On the other hand, there are some signs of disciplinary problems which have a profound negative effect on student outcomes. None of the prioritised reforms – neither the structural one (6-year schools) nor the curricular one (integrated sciences) – seems to have had an effect on student outcomes in science (Table 9.5).

In spite of satisfactory levels of implementation of most recommendations of the reforms, the no-effect findings, and inconsistent relation with science achievement of some of the recommended policies of the program, call for more research to assess the validity of the program's recommendations.

Table 9.5 Student, Teacher, School and Instruction variables, 1994, 1992 and 1995

	1984 %	1992 %	1995 %
Second-generation Israeli students	43	60	66
Students born in Israel	92	90	88
Pupils liking science a lot	25	30	43
Students who perceive themselves doing well	12	18	51
Teachers with an academic degree	4	22	32
Schools where teachers devote <3 hours per week to professional development	50	40	90
Instruction is teacher led (in part, most lessons)	31	54	42
Students copy from board (in part, most lessons)	53	68	81
Teacher demonstrates in laboratory (in part, most lessons)	29	8	35
Students do experiments on their own (in part, most lessons)	39	3	6
Students work in paris/groups without teacher assistance	84	87	24
Schools in which computers are used (often/sometimes)	3	12	33
Schools lacking substantial laboratory equipment/materials	45	38	30
Schools providing after school enrichment activities	26	17	13

Taking into account the minor changes in students' input characteristics over the years, while most of the recommended processes were implemented, a serious question remains regarding the low achievements of Israeli students in the TIMSS study. The Tomorrow 98 program does not seem to have advanced the medium and even lower achievement ranks of elementary Israeli pupils in international comparisons. Why is this? What actually goes on in the classrooms? How do teachers interpret new pedagogies aimed to foster student meaning making (i.e. constructivist pedagogies) and calls to foster meta-cognitive and reflective learning? How do they cope with the call for integrated science teaching? All these questions deserve a more qualitative look at science classrooms.

NOTE

1. This expenditure as a percentage of the Gross National Product rose from 8.5 per cent in 1984 to 9.5 per cent in 1995 and puts Israel above some of the leaders in educational achievement (Australia, England, Korea, the Netherlands) (TIMSS, 1997).

REFERENCES

Baker, P. D. (1997) Good news, bad news, and international comparisons: Comment on Bracey. *Educational Researcher, 26* (3), pp. 16-17

Ben-Peretz, M., and Zeidman, A. (1986) Three generations of curricular development in Israel. *Studies in Education,* 43/44, pp. 317-327

Bracey, G. W. (1996) International comparisons and conditions of American education. *Educational Researcher, 25* (1), pp. 5-11

Bracey, G. W. (1997) On comparing the incomparable: A response to Baker and Stedman. *Educational Researcher, 26* (3), pp. 19-25

Chen, D., and Novick, R. (1987) The challenge is scientific and technological literacy for all. *MABAT Publication, 5.* School of Education, Tel Aviv University (Hebrew)

Chen, M., Lewy, A., and Adler, H. (1978) *Processes and outcomes of educational practice: Assessing the impact of junior high school.* School of Education, Tel Aviv University and the NCJW Research Institute for Innovation in Education, School of Education, The Hebrew University, Jerusalem (Hebrew)

Dan, Y. (1983) The process of becoming a teacher education college. *Mahalachim: The Levinsky Teachers College Annual,* pp. 22-32 (Hebrew)

Etzioni Committee (1979) *The Report of the Committee Appointed to Assess the Status of Teachers and Teaching.* Jerusalem: Ministry of Education and Culture. Also published in *Hed Ha'chinuch,* 55 (18), pp. 7-22 (Hebrew).

Goldstein, H. (1996) Introduction in general issues. The IEA Studies Assessment in Education. *Principal Policy and Practice, 3* (2), pp. 125-128

Israel Believes in Education (1996) Jerusalem: Ministry of Education, Culture and Sports

Keeves, J.P. (1992) *Learning science in a changing world: Cross-national Studies of Science Achievement: 1970-1984* (The Hague: IEA)

Lewy, A., and Chen, M. (1974) *Closing or widening of the achievement gap: A comparison over time of ethnic group achievement in the Israel elementary school.* Research Report No. 6, School of Education, Tel Aviv University (Hebrew)

Martin, M. O. Mullis, I. V. S., Beaton, A. E., Gonzalez, E. J., Smith, T. A. and Kelly, D. L. (1997) *Science achievement in the primary school years* (Chestnut Hill, MA: TIMSS International Study Center, Boston College)

Peled, A. (1976) *The education in Israel during the 80s.* Ministry of Education and Culture, Jerusalem (Hebrew)

Schmida, M. (1987) *Equality and excellence: Educational reform and the comprehensive school* (Ramat Gan: Bar-Ilan University Press [Hebrew])

Shavelson, R., McDonnell, L., Oakes, J. (eds.) (1989) *Indicators for monitoring mathematics and science education: A sourcebook.* (Santa Monica, CA: Rand)

Stedman, L.C. (1997) International achievement differences. An assessment of new perspective. *Educational Researcher, 26* (3), pp. 4-17

Tomorrow 98 (1992) *The report of the high commission of scientific and technological education* (Ministry of Education and Culture [Hebrew])

Zuzovsky, R. (1987) *The elementary school in Israel and science achievement.* Ph.D. thesis. (Jerusalem: Hebrew University [Hebrew])

Zuzovsky, R. (1993) *Supporting science teaching in elementary schools within the policy of extended learning day.* Implementation Study (Center for Curricular Research and Development, School of Education, Tel Aviv University [Hebrew])

Zuzovsky, R. (1997) *Science in the elementary schools in Israel.* Findings from the 3rd International Mathematics and Science Study and a complementary study in Israel. (Center for Science and Technology Education, Tel Aviv University [Hebrew])

Zuzovsky, R., and Aitkin, M. (1993) *Final report: An indicator system for monitoring school effectiveness in science teaching in Israeli elementary schools.* Israel Foundation Trustees Grant 90 and the Chief Scientist's Office (Jerusalem: Ministry of Education and Culture [Hebrew])

APPENDIX 1

Correlation Co-efficients of Students' Background Variables with Science Achievement and Mean Achievements of Sub-populations Defined According to Categories of Each Variable and Increment Above or Below Standardised Mean

	1984 n = 2216			1992 n = 3525			1995 n = 1908		
	Corr. Coeff.	Mean (SD)	Incre.	Corr. Coeff.	Mean (SD)	Incre.	Corr. Coeff.	Mean (SD)	Incre.
Size of family	-.21**			-.12**			-.09**		
Small family		64 (17)	+3		54 (16)	+1		65 (18)	+1
Medium family		61 (17)	+1		53 (16)	+1		65 (17)	+1
Large family		53 (17)	- 4		48 (15)	- 3		59 (17)	- 2
No. of books:	.20**			.20**			.12**		
Few		50 (17)	- 5		46 (15)	- 4		60 (17)	- 1
Up to 100		57 (17)	- 1		52 (15)	0		64 (17)	+1
100 - 200		60 (16)	0		54 (16)	1		67 (16)	+2
More than 200		63 (17)	+2		56 (16)	+3		66 (17)	+2
Born in Israel	.06**			.15**			.07*		
No		57 (17)	- 2		45 (17)	- 4		60 (19)	- 1
Yes		60 (17)	0		53 (16)	+1		64 (17)	+1
No. of generations in Israel	.18**			.18**			.10**		
0		57 (2)	- 2		45 (18)	- 4		60 (19)	- 1
First		57 (17)	- 2		51 (15)	+1		62 (18	- 1
Second		63 (16)	+2		55 (16)	+1		65 (17)	+1
Self concept as science learner	-.00			.10**			.09*		
Low		56 (18)	- 2		45 (16)	-4		57 (19)	- 3
Medium		61 (17)	+1		53 (16)	+1		65 (17)	+1
High		56 (18)	- 2		53 (15)	+1		65 (17)	+1
Attitude toward study of science	.02			.14**			.02		
Does not like		56 (17)	- 1		47 (15)	- 3		63 (17)	0
Likes		60 (17)	0		52 (16)	0		64 (17)	+1
Likes a lot		60 (18)	0		54 (16)	+2		65 (17)	+1

Corr. Coeff. = Correlation coefficient

Incre. = Increment T-score – Points above or below standardised mean

APPENDIX 2

Changes (1984-1995) in Selected Indicators of Teachers' Background Variables

	Per cent of Teachers		
Indicators of Teachers' Background Variables	1984 n = 78	1992 n = 74	1995 n = 75
Two years seminar	36	19	7
Three years seminar	37	53	60
Academic degree	4	22	32
Combined academic and pedagogical preparations	4	11	25
Experienced teachers (more than 20 years)	17	12	21
Teacher's professionalism: a) Per cent of schools where teachers devote more than 3 hours per week to extra teaching activity (Planning Act 3)	87	87	77
b) Per cent of schools where teachers devote more than 3 hours per week to professional development activities (in-service) (Act 6)	50	40	70

APPENDIX 3

Changes (1984 - 1995) in Structural Variables of Schools

	Per cent of Schools		
Indicators of School Type (Variable category used in brackets)	1984 n = 70	1992 n = 74	1995 n = 50-52
6-year school (1)	50	66	77
Small school (1)	67	63	54
Inner-city school (3)	72	–	42

APPENDIX 4

Correlation Coefficient of School Structural Variables with Science Achievement and Achievement Means of Sub-populations Defined According to Variable's Categories

	1984 n = 70			1992 n = 74			1995 n = 52		
School Structural Input Variables	Corr. Coeff.	Mean (SD)	Incre.	Corr. Coeff.	Mean (SD)	Incre.	Corr. Coeff.	Mean (SD)	Incre.
Six or eight-year school	-.05			-.34**			-.00		
1 = six		60 (13)	0		43 (7)	+1		62 (7)	0
2 = eight		59 (9)	-1		48 (8)	-2		62 (6)	0
School size	.16			.02			-.05		
1 = small		67 (10)	-1		55 (8)	0		63 (7)	0
2 = medium		60 (10)	+1		53 (6)	0		62 (7)	-1
3 = large		59 (15)	+1		49 (5)	0		68 (3)	+3
School/Community	-.09			-			.13		
1 = rural		61 (5)	+1		-	-		62 (7)	0
2 = suburb		61 (14)	0		-	-		61 (6)	-1
3 = inner city		59 (10)	-1		-	-		64 (5)	+1

Corr. Coeff. = Correlation coefficient

Incre. = Increment T-score - Points above or below standardised mean

APPENDIX 5

Changes in Percentage of Schools with Inadequate Conditions that Constrain the Implementing of Science Curriculum. Correlation Coefficient of this Variable with Science Achievement and Means of Science Achievement of Sub-populations Defined According to Level of Inadequacy with Increment T-Scores

School conditions	1984 n = 74				1992 n = 71				1995 n = 50			
	% of schools	Corr. Coeff.	Mean (SD)	Incre.	% of schools	Corr. Coeff.	Mean (SD)	Incre.	% of schools	Corr. Coeff.	Mean (SD)	Incre.
Schools with inadequate conditions	45	-.21			38	-.27*			30	.09		
Adequate conditions (not constrained)			67 (11)	+4			55 (7)	+2			63 (6)	0
Partially adequate (partially constrained)			60 (9)	0			53 (8)	0			61 (7)	-1
Highly inadequate (highly constrained)			59 (8)	-1			49 (7)	-2			64 (6)	+1

Corr. Coeff. = Correlation coefficient

Incre. = Increment T-score - Points above or below standardised mean

APPENDIX 6

Changes (1984-1995) in Frequency of Indicators of School Level Processes

Indicator of School Level Processes (Category used in brackets)	Per cent of Schools		
	1984 n = 73	1992 n = 75	1995 n = 50
Implementing the 'new' or school with own science (1 – yes) curriculum (2 – full implementation)	71	65	36
Integrating science with other school subjects (0 - not integrated; 1 – integrated)	–	24	1
Provision of enrichment programmes (science clubs, projects, enrichment lessons) (1 – yes; 2 – no)	26	17	48
Out of school science activities – field trips (1 – yes)	96	92	–
Actual use of computers in teaching and learning - per cent of schools where computers are used often (1 – often; 2 – sometimes)	3	1	1
Administration deals with discipline problems in school (3 – often; once a week, once a day)	30	9	27

APPENDIX 7

Correlation Coefficient of School Processes with Science Achievement and Breakdown of Sub-populations Mean by Variable Categories with Increment T-Scores

Indicators	1984			1992			1995		
	Corr. Coeff.	Mean (SD)	Incre	Corr. Coeff.	Mean (SD)	Incre	Corr. Coeff.	Mean (SD)	Incre
Implementing 'new' or additional school developed science curriculum	.23			.40***			.25		
1 = not at all, part		56 (11)	-2		47 (7)	-3		61 (6)	-1
2 = full implementation		61 (9)	+1		54 (7)	+1		64 (6)	+1
Integrating science with other subjects	-			-.08			-.00		
0 = not integrated		-			52 (9)	0		63 (6)	0
1 = integrated		-			51 (8)	-1		62	-1
Enrichment (science clubs, project, enrichment lessons)	-.0			-.00			-.18		
1 = yes		60 (10)	0		51 (9)	0		64 (6)	-1
2 = no		59 (10)	0		51 (7)	0		61 (7)	0
Out of school (field trip) activities	-.09			-.11			-		
1 = yes		59 (10)	0		52 (7)	0		-	-
2 = no		55 (6)	-3		48 (9)	-2		-	-
Use of computers in classroom	-.10			.04*			-.03		
1 = often		65 (5)	+3		49	-1		65	+1
2 = rarely		60 (9)	0		52 (7)	+1		63 (7)	0
Disciplinary problems (class disturbance)	.04			.17			-.29*		
1 = seldom		57 (9)	-1		48 (9)	-2		65 (5)	+1
2 sometimes		59 (9)	0		52 (7)	0		61 (6)	-1
3 = often		59 (13)	0		52 (9)	0		61 (8)	-1

Corr. Coeff. = Correlation coefficient

Incre. = Increment T-score – Points above or below standardised mean

APPENDIX 8

Per cent of Schools where Type of Instruction Occurs Often

	Per cent of Schools		
	1984	1992	1995
Indicators of Whole Class Teacher-Led Instruction:			
Copying from board (1-3) 1 = often	53	68	81
Teacher checks homework (1-3) 1 = often	31	10	67
Teacher shows how to solve problems (1-3) 1 = often	–	9	23
Teacher demonstrates (1-3) 1 = often	29	8	35
Teacher answers questions in most and every lesson 3+4 (1984, 1992)	84	86	43
Whole class teacher led instruction in most and every lesson (1995)	16	14	57
Teacher lectures (1984 and 1992) and whole class teacher led instruction (1995) in most and every lesson 3+4	31	54	42
Indicators of individualised instruction:			
Do experiments (1-3) 1 = often	39	3	6
Work individually without assistant (1-4) (in most and every lesson – 3+4)	39	58	21
Indicators of group work:			
Work in pairs or small group (1-3) 1 = often	36	9	15
Work in pairs without assistance (1-4) (in most and every lesson – 3-4)	84	87	24

APPENDIX 9

Correlation Coefficients' of Instructional Variables with Science Achievement and Breakdown of Sub-population Mean Achievement Score by Variable Categories

Indicators	1984 n = 83			1992 n = 78			1995 n = 82		
	Corr. Coeff.	Mean (SD)	Incre	Corr. Coeff.	Mean (SD)	Incre	Corr. Coeff.	Mean (SD)	Incre
Teacher Led Instruction									
Copying from board (Dichotomised av. 1-3)	-.01			-.16			.23*		
1 = frequent		59 (10)	0		53 (8)	1		62 (6)	0
2 = infrequent		59 (9)	0		50 (9)	-1		66 (7)	+2
Teacher checks homework (Dichotomised av. 1-3)	-.07			-.09			-.09		
1 = frequent		60 (8)	0		54 (8)	+1		63 (6)	0
2 = infrequent		59 (10)	-1		51 (9)	0		62 (7)	0
Teacher demonstrates in lab (Dichotomised av. 1-3)	.19			-.21			-.19		
1 = frequent		57 (8)	-2		58 (16)	+4		65 (5)	+1
2 = infrequent		60 (10)	0		51 (7)	0		62 (7)	0
Teacher answers questions (Dichotomised var. 1-4)	-.12			-.08			.18		
1+2 = infrequent		63 (3)	+2		53 (3)	+1		60 (8)	-1
3+4 = frequent		59 (8)	-1		51 (7)	-1		64 (5)	0
Teacher lectures (Dichotomised var. 1-4)	-.21			-.00			.18		
1+2 = infrequent		61 (7)	+1		51 (7)	0		60 (8)	-1
3+4 = frequent		56 (13)	-3		51 (7)	0		64 (5)	+1
Individual Instruction									
Do experiments (Dichotomised av. 1-3)	-.21			-.19			-.23*		
1 = frequent		62 (11)	+1		61 (1)	+6		69 (3)	+4
2 = infrequent		58 (9)	-1		61 (8)	0		63 (7)	0
Work individually without assistance (Dichotomised var. 1-4)	.24*			.28*			.13		
1+2 = infrequent		59 (7)	-2		49 (7)	0		61 (6)	-1
3+4 = frequent		61 (7)	+2		54 (12)	+1		63 (5)	+1
Group Work	.00			-.25*			.29**		
Work in pairs, small groups (Dichotomised av. 1-3)									
1 = frequent		59 (10)	0		58 (6)	+4		58 (6)	-3
2 = infrequent		59 (10)	0		51 (8)	0		64 (7)	+1
Work in pairs (Dichotomised var. 1-4)	.14			.19			.07		
1+2 = infrequent		58 (8)	-1		47 (6)	-3		62 (7)	0
3+4 = frequent		60 (9)	0		52 (7)	0		63 (5)	0

Corr. Coeff. = Correlation coefficient.   Incre. = Increment T-score – Points above or below standardised mean

[1]   The correlation coefficients here are of dichotomous variables

Chapter 10

The need for caution when using the results of international mathematics testing to inform policy decisions in education in the member countries

Ann Kitchen
Centre for Mathematics Education, University of Manchester, UK

1. INTRODUCTION

The Third International Mathematics and Science Study (TIMSS), carried out in 1995, tested students within three age-groups in many countries (see Chapter 2). While its aims were varied and looked at what students are intended to learn, who provides the instruction, how instruction is organised and what students have learned, the media and the politicians in England have focused solely on the data on student achievement. However, it is the misleading interpretation given to the results and the use to which this has been put so far that are disturbing. England and Scotland, both of which took part in the study of the two younger age-groups, appeared to do worse in mathematics than many of their European partners and significantly worse than those from the Pacific Rim. Is this true? Do students in England at ages nine and 13 know less mathematics? This chapter will look at the study design and the data for Population 2 students from the two grades containing the highest proportion of 13 year-old students. However many of the conclusions can equally apply to Population 1.

TIMSS has been scrupulous in ensuring that data are available for research purposes. However, many of the findings were made public before the main data were published and it is these that have been seized on. Headlines such as 'England lags behind the continent in Mathematics', 'What is wrong with our mathematics teachers?' together with a call to ban

D. Shorrocks-Taylor and E.W. Jenkins (eds.), Learning from Others, 219–231.
© 2000 *Kluwer Academic Publishers. Printed in the Netherlands.*

calculators have been based on a misunderstanding of the TIMSS results. Why did this occur? The summaries were open to misinterpretation by those who had not been involved in the survey. Thus the first summary for England stated, 'The first National Report on TIMSS Part 1 compares the performance of 13 year-old students in England on the TIMSS mathematics and science tests with that of students of a similar age in other countries.' (NFER, 1996). A paragraph in the same document stated:

> Students in England achieved relatively high mean scores in science but relatively low mean scores in mathematics.

The word relatively was soon missed out. This led to the misconception that the data showed that student performance in England in science was better than that in mathematics. This has not been supported by the data. There has been no work done yet to quantify the difference in difficulty of the two sets of questions and it is entirely possible that English students' performance is the same in science and mathematics but in many other countries, where students do not study science to the same depth, the science results were considerably worse.

2.   THE MISCONCEPTIONS ARISING FROM PUBLICATION OF THE TIMSS SUMMARIES

One of the main purposes of the TIMSS study would seem to be to collect international data to see how the teaching and learning of science and mathematics might be improved by looking at examples of good practice from countries which had achieved well on the tests. This is certainly the case as far as England is concerned. Politicians and education officials, as well as the general public, have taken England's position in the league tables as showing which countries should be emulated if England is to improve its position in the league tables. While the rankings certainly give some information about relative achievement, one of the major benefits of the study, the production of data on the relation between questions English students found difficult and the other data collected from teacher and pupil questionnaires has, by and large, been ignored

Why is it not possible to take England's ranking in each of the tables as a direct indicator of its position relative to other countries? What other factors must be understood before any attempt to interpret the data can take place? Firstly, the limitations of who was tested, how the testing was carried out and how the tests were marked and analysed must be taken into account. In the year following the publication of the Population 2 results, over a hundred

10. The need for caution when using the results of international   221
mathematics testing to inform policy decisions in education in the
member countries

teachers were asked the following questions in relation to their understanding of the TIMSS tests.

1. There were two tests, one maths and one science.
   True / False
2. English students did better on the science test than the maths. test.
   True / False
3. Which of the following is true?
   The '13 year-old' cohort tested were:
     a) those 13 year-olds in a specified year;
     b) those in the year with the largest number of 13 year-olds;
     c) those in the two consecutive years with the largest number of 13 year-olds;
     d) ranged in age from 12 to 15;
     e) all countries had a mean student age of between 13 and 14;
     f) the same in all countries.
4. There was a mixture of multiple-choice and open-ended questions.
   True / False
5. Students used to using a calculator were not disadvantaged.
   True / False

The teachers were surprised to hear that students worked through a set of booklets and within those booklets, the questions were mixed up both in difficulty and in subject. Thus a single booklet would have a mixture of science and mathematics questions. They all thought that the difference in performance in mathematics and science in English schools as measured by performance against that of the international community meant that science teaching in English schools was more successful than mathematics teaching.

As far as the question on the cohort tested was concerned, once again teachers agreed that the impression that they had gained from the press was that all 13 year-olds were tested. Most teachers were aware of the style of questioning used and a large number had seen some of the published questions. All agreed that those they had seen were appropriate, although some misgivings were raised about multiple-choice questions in general and the wording of some questions in particular. All teachers felt that those students used to using a calculator in Population 2 would be disadvantaged by not having a calculator for many of the mathematics questions, in particular number, percentages and fractions, but not for the science questions. It was also felt that a true test of mathematical ability, as opposed to numeracy, should include some questions where the use of a calculator was assessed.

3.   PERCEIVED DRAWBACKS
         IN THE STUDY DESIGN

In carrying out any international study, a number of assumptions must be made in order to make comparisons. What these assumptions are, and how valid they are, would appear to have a direct impact on the way the results can be interpreted. If a set of league tables were to be produced then the implicit assumptions which must be satisfied would appear to be as follows.
1. The cohorts chosen are the same in each country.
2. The types of question used in assessment are equally familiar to students from all countries.
3. The questions chosen are representative of the curricula of all the countries.
4. The two subjects are tested independently.

It is necessary to look at these assumptions together with others it appears were made by the TIMSS team.

3.1   The cohorts chosen are the same in each country

The study took a sample from the two adjacent classes that contained most children of the required age. However, this meant that the actual age of the children studied in different countries was very varied. In the main reports, the league tables were produced for each of the adjacent classes, (called seventh and eight grades). However this meant that in some countries the children tested ranged in age from eleven to 15. The distribution of children at each of the notional ages was given by TIMSS from which is constructed the approximate age distribution of the children assessed for a group of countries (Figure 10.1).

What does this distribution show? Firstly, that in addition to the class with most of the 13 year-olds, Scotland tested the class with mainly 12 year-olds and a few 13 year-olds, while Japan tested the class with mainly 14 year-olds together with a few 13 year-olds. France belongs, like most of its European partners, to a totally different school system. As can be seen from the chart, only 80 per cent of the 13 year-olds were tested, while 20 per cent of the 15 year-old cohort were tested. Why was this? In France, students who have not reached an acceptable standard will not progress to the next grade. Thus the 15 year-old students are those who have taken longer to reach the standard needed for that grade. The missing 13 year-olds are those who have not yet reached the required standard while some of the 12 year-olds are those who have achieved the standard early and have moved to the next grade early. There were even a few 11 year-olds. It appears that there are two very different cohorts here and league tables encompassing both

10. The need for caution when using the results of international   223
mathematics testing to inform policy decisions in education in the
member countries

systems are very suspect. Table 10.1 shows some of the countries belonging
to these two systems.

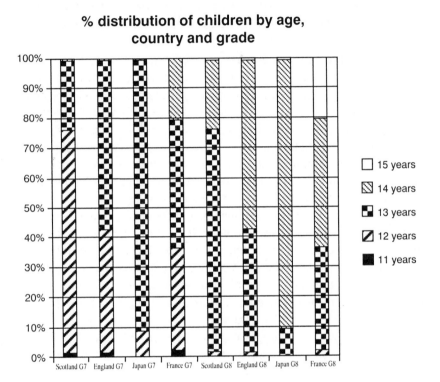

**% distribution of children by age,
country and grade**

Figure 10.1. Age distribution of children for Population 2

Table10.1. Countries by type of grade cohort

Countries with students allocated to grades purely by age:						
Japan	Sweden	Scotland	Norway	New Zealand	England	Iceland

Countries with students allocated to grades by achievement:						
Hong Kong	Switzerland	Belgium	Czech Rep	Hungary	Canada
Ireland	USA	Latvia	Spain	Portugal	Austria
Netherlands	Germany	Romania	Thailand	South Africa	France

(TIMSS Tables 1.8 and A3 Mathematics achievement in the middle years, Nov 96)

The statistics in the main reports are given separately for both of the two
adjacent year groups so it is possible for countries within the same education
type to adjust for inequities in the age of their samples. However this is not
possible between the two types. It would have been much more appropriate

to test all 13 year-olds, no matter what grade they were in, if meaningful comparisons between student performance from one country to another were to have been made.

3.2   The types of questions used in the assessment are equally familiar to students in all countries

This was probably impossible to achieve. The tests were mainly multiple-choice chosen partly for its ease of marking and relative cheapness. There were smaller numbers of open response and extended questions. In addition, students from some countries took performance assessment tasks but these were reported some time after the results of the written tests and were never aggregated with them. This was probably because their cost meant that many countries did not take part in the tasks. Great care was taken in translating the tests into many different languages but even then the language available was not always the native language of the students taking the test. This was more of a problem in some countries than others.

There has been much discussion on the appropriateness of the multiple-choice format. It can be said that it leads to students being trained to spot which is the correct answer when a set of possible answers is given, rather than being able to solve the problem themselves. Many students using the multiple-choice format, especially for mathematics, do not use the expected content knowledge to get the right answer. In many cases, a lower level skill will enable the possible choices to be reduced from four or five to two thus enabling the student to have a fifty-fifty chance of getting the correct answer. As a diagnostic tool it is equally open to question. It may be that those choosing an answer that shows a particular form of mathematical misconception do so, not because they hold that particular misconception but because it happened to be chosen at random. A third drawback is that in many cases, distracters, answers that seem wholly plausible, tempt students to choose them, when by working out the answer from first principles the error would have been avoided.

While these are general shortcomings to all multiple-choice tests, the actual questions are very difficult to write effectively and it can be argued that in the case of TIMSS some questions fell short of the required standard. The following, drawn from the tests, exemplify these types.

Type 1:   Can be solved by lower level skills:

A straight line passes through the points (3,2) and (4,4).
Which of these points also lies on the line?
A (1,1)   B (2,4)   C (5,6)   D (6,3)   E (6,5)

10. The need for caution when using the results of international   225
mathematics testing to inform policy decisions in education in the
member countries

There was no strong distracter here. Interestingly the percentage of correct answers for students in Population 2 from England was higher than that for all countries except for the Netherlands, while Scotland was not far behind. The correct answer (5,6) was chosen by 56 per cent of the students from England. Given that there is no obvious distracter, it implies that 45 per cent of the students actually knew the correct answer and eleven chose it at random. The item was designed to be one of the geometry items but it is possible that many students simply saw it as a number pattern pair and counted on 3,4,5 for the first number and 2,4,6 for the second without understanding the relevance of the co-ordinates or the equation of a line.

Type 2:   Where the language medium could affect the strength of the distracter:

If the price of a can of beans is raised from 60 cents to 75 cents, what is the per cent increase in the price?
A 15%   B 20%   C 25%   D 30%

The distracter A, (where children confuse cents with per cent) was probably greater for those countries where the currency was left unchanged.
The percentage of correct answers here was interesting. Many countries had percentages lower than 20 per cent. If students chose at random, 25 per cent of the answers would be expected to be correct. This is a sign of a strong distracter.

Type 3:   Where because of faulty question wording more than one of the answers are correct but only one is allowed:

Whenever scientists carefully measure any quantity many times, they expect that

A   all of the measurements will be the same,
B   only two of the measurements will be the same,
C   all but one of the measurements will be the same,
D   most of the measurements will be close but not exactly the same.

The 'correct' answer (D) would be obtained by those students who were thinking of measurement with a continuous scale, the length of a piece of wood, the time taken for a ball to roll down a slope. However those thinking of an experiment where the result to be measured consisted of discrete data, the number of peas in a specified packet, would correctly expect to get the same number no matter how many times they were counted. Those reading it as a probability experiment, 'How many times would a head be thrown after a hundred tosses?' would correctly accept that the number thrown could vary quite widely each time it was done. Only the first answer described gained any credit. This is a question that, by not defining the type of measurement required, has allowed thinking students to be penalised. Unfortunately the percentage data for each choice are not available here, only the percentage getting the question 'right'.

These last examples have looked at the problems with multiple-choice questions. What about the free response questions? These are a more reliable way of testing knowledge. However, they are more expensive to mark reliably and in some cases can lead to a student failing to get credit because the answer given, although valid, does not touch upon the particular piece of knowledge the question was designed to elicit. Secondly it is easy, especially in science, for answers to be too simplistic.

Unfortunately, the method of standardising the questions meant that a question with a low international score was deemed to be harder than one with a higher international score and given a higher weighting in the overall score. This led to unsatisfactory questions being weighted highly.

4.   THE QUESTIONS CHOSEN WERE REPRESENTATIVE OF THE CURRICULA OF ALL THE COUNTRIES

Again much care was taken with this. However some compromises had to be made if the questions were not to be too limited. In order to guard against any unfairness, the countries concerned were asked to choose a subset of questions that were particularly suitable to their own students and a Table of each country's score on these items alongside those of others was produced. In many cases this showed very little difference to the rank order using the whole bank of questions. However, this is linked to another major assumption that has been made by the TIMSS researchers, namely that in a study of this sort it is appropriate to ban calculators from the mathematics tests. If the researchers are testing students' ability to do arithmetic without any aids to calculation then it is an appropriate assumption. However, this was not the case. The reason for banning calculators is evident. While it would be impossible to include questions requiring a knowledge of calculator skills in a test where many of the participating students had not

10. The need for caution when using the results of international   227
mathematics testing to inform policy decisions in education in the
member countries

met a calculator, it is possible to restrict the test to questions that do not require any computational aids and expect all the students to be able to attempt them. However, it is open to question whether this leaves the students on any more of a level playing field. Those students who have been used to using a calculator will in many cases have not refined their number handling skills, especially with fractions. In addition, even when such students have the required skills their speed at such calculations may be much reduced. Thus while the rankings may be a fair assessment of numeracy, they may not be true of mathematics in a wider sense.

5.   THE TWO SUBJECTS WERE TESTED INDEPENDENTLY

The need that this implied, for two distinct timed tests, was not acknowledged by the TIMSS researchers.

They felt that the two sets of questions could be separated after the event. It is open to question whether this is appropriate. Certainly many would say that by mixing the two sets of questions together it became impossible to remove the effects of one on the other. Did those who could not do the science questions spend longer on the mathematics? If they did, does it matter?

6.   OTHER QUESTIONS ON DESIGN THAT MAY AFFECT THE INTERPRETATION OF THE RESULTS

6.1   The data about classroom practice can be used to explain the students' performance

Understanding of this point is especially important when looking at calculator usage. This is shown clearly when analysing the data on students' use of calculators in relation to their tests scores.

Figure 10.2 shows the teachers' reports of frequency of use of calculators in mathematics classes, by mean achievement in mathematics test for the Upper Grade of Population 2.

What conclusions can be drawn from this table? Firstly it suggests that within a country, the more frequently a calculator is used, the higher the students achievement in mathematics. However, it might equally mean that in a certain country students were only allowed to use a calculator once they had reached a certain standard of attainment. Even then, if this is not the case, the Figure itself does not show how long the students concerned have been using the calculator. Only if the use of calculators has been taking place for some time might these two factors be related. As far as England is concerned, the data do support the view that appropriate calculator usage need not lead to lower performance.

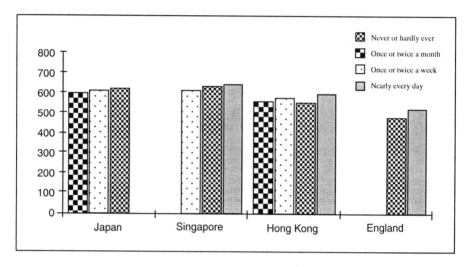

Figure 10.2. Teachers' Reports of Frequency of use of Calculators in Mathematics Classes by mean achievement in Mathematics test: Upper Grade of population 2

6.2   The scores given can be taken to show raw differences in achievement

A different misunderstanding can have more serious effects. This is that the score in the league tables, given out of 800, has any simple relationship with the number of questions answered. The lack of such a relationship might seem obvious to a statistician who has read the report carefully, but to most people a mean score of 400 meant that the students got an average of 50 per cent of the questions right and students from a country with a mean score of 480 answered an extra 10 per cent of the questions correctly. This is not the case. Because of the number of items tested, any one student only did a subset of them. It was therefore decided to standardise the scoring so that a student who had difficult questions was given more credit than one who answered an easier selection. The choice of the weighting factor for a question was decided, in part, by the number of students answering it

10. The need for caution when using the results of international   229
mathematics testing to inform policy decisions in education in the
member countries

correctly. The drawback here is that an unsuitable question, answered badly, or mis-marked, could have a disproportionate effect on the overall score. In addition, it is impossible for the outsider to make any estimate about the actual validity of the weighting. Another effect of standardising the scores means that an artificial mean and standard deviation are forced onto the data. The decision was to choose the population mean as 500 and the population standard deviation as 100. There is no way of knowing how many correct answers a difference of 16 in the scores of France and England for their 13 year-olds actually signifies. (Table 10.1). The only thing that can be agreed is that on average French students did answer more questions correctly. It is impossible to state from the results whether this was in the order of ten questions or 0.1 of a question. It would seem that for policy makers the distinction between the two is of paramount importance.

6.3   How should this affect the use of the findings?

League tables make compelling reading. It is tempting to look at the summary tables and draw conclusions from them. However, drawing conclusions from them and acting upon those conclusions needs considerable thought.

Firstly, in order to make valid comparisons, the populations to be compared must be the same. The results for the seventh grade students in Japan, England and Norway can be compared, but there is a case for comparing these with the students from the eighth grade in Scotland given the disparity in the age cohort. (This is shown clearly in one of the few Tables giving the scores for 13 year-olds, where Scotland had a higher score than England).

The results for countries such as France and Germany with their different sampling cohorts should not be taken as a true comparison with countries of the other sampling type and, while large differences in scores may indicate a different level of achievement in the populations as a whole, small differences in the individual scores cannot be said to be significant.

If comparisons are to be made, it is possible to do so by looking at the summaries of median scores for those 13 year-olds in the survey.

While it is tempting to draw conclusions from the published results of the questionnaires on student and teacher classroom practice in the grades studied, it must be realised that in many cases the outcomes perceived as far as attainment is concerned will be the result of practices from earlier in the students' school career. This is especially true in the case of calculator usage. It is more informative to find out when calculators were introduced to the curriculum of the students studied rather than whether the students use them frequently at the present time. Certainly given the rate of change in the education system, information on the present practice for younger cohorts cannot be used to inform what happened to an older cohort when they were of a similar age.

Given the variability of the questions and marking, it is more appropriate for ranking on each item to be studied. This has the advantage that an informed decision can be made on the suitability of that item. A disadvantage here is the small number of students taking each individual item. While the sample as a whole ran to several thousand for each country, the numbers taking each item were much lower, in many cases less than 300.

6.4   What duty does the researcher owe the wider community?

It is the responsibility of any researcher to carry out a study in a responsible way and publish the results fully. Does the responsibility end there? It can be argued that the academic community has a duty to monitor the use that such data are put to and where it sees that the results of such work is being misinterpreted it should both publish a correction and take steps to see that future work does not lead to similar misconceptions.

It is pleasing to note that the forthcoming OECD study of student performance is taking account of many of the factors described here. So, for example, they will be testing year cohorts rather than grade cohorts and there is hoped to be less reliance on multiple-choice items.

However, a great deal needs still to be done in monitoring and correcting the public and governmental misconceptions that may be produced from the publication of research work. The academic community of educational researchers owes this duty to the community at large.

6.5   What caution should be exercised by those using the results of published studies such as TIMSS?

It is vital that care is taken to wait for the full, published data rather than relying on summaries. In addition, if appropriate action is to be taken it is vital that both the design and the results are scrutinised carefully. In many cases the raw data will give a better picture of the actual relationship than the standardised data. In addition, care must be taken not to draw spurious conclusions from apparent relationships. It is all too easy to think that because two factors appear to have some relationship, one causes or is caused by the other. They may both be caused by an as yet untested factor, or the relation may be due to some quirk of the sampling process.

10. The need for caution when using the results of international   231
mathematics testing to inform policy decisions in education in the
member countries

International studies can be of immense use in evaluating the educational system of an individual country but caution must be observed both by researchers and those who use the results of their labours. No decisions should ever be made on the basis of the summaries of research findings. Only after serious consideration of the initial data and the research design should changes be made. Even then, it is vital that politicians, civil servants and academics alike realise that changes, made with the best of intentions, often have a result opposite to that intended.

NOTES

The full TIMSS reports on which this chapter based, together with the original tabulated data, are available on the web. http://wwwcsteep.bc.edu/timss

Chapter 11

International Comparisons of Pupil Performance and National Mathematics and Science Testing in England

Diane Shorrocks-Taylor, Edgar W Jenkins, Janice Curry, Bronwen Swinnerton
School of Education, University of Leeds, UK

1. THE BACKGROUND TO THE STUDY

It is fair to say that the results of the 1995 TIMSS work proved something of a disappointment to both educationists and policy makers in England. In broad terms, in the 'Multiple Comparisons' analyses for mathematics, the English sample scores placed them 17th out of 26 in the Population 1 comparisons (Year 5) and 25th out of 41 in the Population 2 comparisons (Year 9). In science, the results were a little different, with the English sample being placed 8th out of 26 in the Population 1 (Year 5) comparisons and 10th out of 41 in those for Population 2 (Year 9). The results for Year 4 and Year 8 were broadly similar. All figures used here are drawn from Keys *et al.*, (1996) and Harris *et al.*, (1997). The main findings as reported are summarised in more detail in Table 11.1 below, and further information is available in Keys (1999).

Perhaps even more significantly, in the 'Test-curriculum matching' analyses (Keys *et al.*, 1996 and Harris *et al.*, 1997), an equally complex picture emerged. In these analyses, each country selected the questions most closely related to its curriculum and comparisons were made across several countries in terms of only these selected items. In mathematics, performance on this specially selected sub-set deemed most relevant to the mathematics curriculum in England showed the English sample to emerge as 8th out of the nine countries included in the Population 1 analysis (Year 5) and 9th out of eleven in the countries included in the Population 2 (Year 9) equivalent investigation. In science, the parallel analyses placed the English sample 6th

D. Shorrocks-Taylor and E.W. Jenkins (eds.), Learning from Others, 233–258.
© 2000 *Kluwer Academic Publishers. Printed in the Netherlands.*

out of nine in the Population 1 (Year 5) comparisons and 4th out of eleven for Population 2 (Year 9).

Table 11.1. Summary of the TIMSS results in England for Population 1 and Population 2

	Mathematics	Science
Population 1	• The scores of the English sample of pupils in mathematics were significantly lower than those of pupils in about half of the countries taking part in the study, and significantly higher than those in about a quarter of the countries. • There were no significant gender differences in the overall mean mathematics scores in this population, as in most of the countries taking part.	• The scores of the English sample of pupils in science were significantly higher than those of pupils in half of the other countries taking part in the study. • Thirteen per cent of pupils in England were in the international top ten per cent in both year groups in science (Years 4 and 5). • There were no significant gender differences in overall mean science scores, although the scores of boys were higher than those of girls in earth science.
Population 2	• The scores of the English sample of pupils in mathematics were significantly lower than those of pupils in about half of the countries taking part in the study, and significantly higher than those in about a quarter of the countries. • The relative position of England in mathematics appeared to have deteriorated slightly since the previous comparative studies. • There were no significant gender differences in the overall mean mathematics scores in this population, except for Year 8 where boys achieved a higher mean overall mathematics score.	• The scores of the English sample of pupils in science were significantly higher than those of pupils in over two-thirds of the other countries taking part. • About one in six of pupils in England were in the international top ten per cent in both year groups in science (Years 8 and 9). • The relative position of England in science appears to have improved since the previous comparative studies. • Boys' overall mean science scores were significantly higher than those of girls in the English sample and this was also the case in about three-quarters of the countries taking part. These gender differences only reached statistical significance in one area, *Chemistry* in Year 8.

In England and Wales, unlike many countries participating in the study, most of what is taught in schools (both primary/elementary and secondary) from the age of 5 to 16 is specified in the form of a National Curriculum. This curriculum covers the core subjects (English, Welsh [in Wales], Mathematics, Science) and the foundation subjects (Technology, History, Geography, Modern Languages, Music, Art and Physical Education) and detailed curriculum specifications are provided in each of these subjects.

The matter does not rest here however, since the National Curriculum also includes a testing programme in the core subjects. All 7 year-olds in

England are tested in English and Mathematics and all 11 and 14 year-olds are tested in English, Mathematics and Science. In the case of the tests for 11 year-olds, the results are collected each year and are published nationally on a school-by-school basis, providing information to the government, policy makers, parents and the wider community. The tests are developed each year and are based directly on the curriculum as specified in the core subjects. Year-on-year equating is achieved through a variety of measures, both quantitative and qualitative. Published results also appear to be showing an improvement in the performance of English pupils, a trend that proved almost in direct contradiction to the TIMSS results. This situation therefore provides an interesting context for further investigation and research.

2.   THE FOCUS AND AIMS OF THE STUDY

The major focus of this chapter is the investigation of the TIMSS results in mathematics and science (for the English sample) in relation to the National Curriculum and its assessment. The research was funded by the Qualifications and Curriculum Authority, the body responsible for the National Curriculum and its assessment in England and Wales. The research agenda was agreed with the Authority and the focus questions were as follows, for both Populations 1 and 2.
1. In mathematics, were the content and skill areas represented in the TIMSS tests and the National Curriculum tests similar?
2. Did the performance of pupils differ in the two kinds of tests in relation to the various content and skill areas in mathematics?
3. In science, were the content and skill areas represented in the TIMSS tests and the National Curriculum tests similar?
4. Did the performance of pupils differ in the two kinds of tests in relation to the various content and skill areas in science?
5. Can the differential performance in mathematics and science in the TIMSS results be explained?
The full results of the study are reported elsewhere (Shorrocks-Taylor *et al.,* 1999) so, for this chapter, the main focus will be on research questions 1, 3 and 5.

3.   THE DESIGN AND METHODOLOGY

In order to be able to compare both the content and skill areas of the two kinds of tests and performance within them, all test questions (TIMSS and National Curriculum tests) were categorised into curriculum and skill areas considered appropriate to both the National Curriculum and the TIMSS tests.

The TIMSS 'Content' areas and 'Performance Expectation' categories were not fully appropriate so it was necessary to provide a framework that would encompass both, in order to provide a metric for comparison. Such categorisations and re-categorisations are not, of course, without their problems: whether in mathematics or science, there are potentially many ways to slice the judgmental cake, and this proved to be the case here. Efforts were made to ensure that the final categorisations in both mathematics and in science reflected some general agreement among the professionals involved in the categorisation process, but it was impossible to ensure that this was total. However, the criteria were applied in a systematic way, the results are therefore defensible and they provide a valid basis for comparing the content and skill areas in the two kinds of tests.

4.   THE CATEGORISATION PROCESS

The process of categorisation of all the questions in both subjects followed several stages, and involved consultation with a wide range of expertise, including Professional Officers at the Qualifications and Curriculum Authority, mathematics and science educators at Leeds University, and primary, middle and secondary school teachers with the relevant subject expertise. The process of categorisation therefore involved the refining of definitions and using exemplification to ensure a standard that could be applied across both TIMSS and National Curriculum questions. A subset of items was also used in a small validation exercise with two independent mathematics and science educators, following the defined criteria, in order to ensure that questions had been allocated systematically across both the TIMSS questions and the National Curriculum questions.

The test material comparisons made in this study are summarised in Table 11.2 below. This shows the dates and National Curriculum tests used in relation to Populations 1 and 2 of the TIMSS study.

Table 11.2 shows that for Population 2, the test comparisons are appropriate. By this we mean that the National Curriculum tests both in 1995 and 1996 were intended for an age group largely similar to those of the TIMSS age cohort. The comparisons for Population 1, however, are somewhat less direct since the TIMSS Population 1 cohort was aged 9 to 10 whilst the National Curriculum tests were intended for 11 year-olds (Year 6). Non-statutory National Curriculum tests are, however, available in both mathematics and science for 9 year-olds in England, and inspection of these revealed broad similarities in terms of the content and skill areas addressed. All this suggests that the Population 1 comparisons to be reported here are valid in terms of the content and skill areas but should be treated with some caution in the context of pupil performance.

Table 11.2. The test comparisons (and timings) in the study

	English school years	**April 1995**	**May 1995**	**May 1996**	**May 1997**
TIMSS *Population 1*	Year 4	TIMSS 1			N/C Y6 tests
	Year 5	TIMSS 1		N/C Y6 tests	
TIMSS *Population 2*	Year 8	TIMSS 2		N/C Y9 tests	
	Year 9	TIMSS 2	N/C Y9 tests		

5. THE TESTS

5.1 The characteristics of the TIMSS tests

The TIMSS tests had the following characteristics.

- Each pupil took two test papers (A and B) each of which contained both mathematics and science questions but with very different balances across each pair of tests.
- The questions in the TIMSS tests were mainly multiple-choice (approximately 75 per cent in all the tests in both Population 1 and Population 2).

5.2 The characteristics of the National Curriculum tests in England

- In mathematics, the tests for 11 year-olds consist of two test papers (A and B), based directly on the curriculum specification. Each paper is 'ramped', in the sense that there are easier questions at the beginning, but they gradually become harder through to the end. The questions are varied in terms of their type (calculation, short answer, diagram completion, explanations etc.) and the extent to which they are set in real-life contexts. For one of the papers (Test B) pupils are expected to use calculators.
- In science, the tests for 11 year-olds also consist of two test papers (A and B) but perhaps because of the more composite nature of the science curriculum, the questions in both test papers are less obviously 'ramped' in difficulty from the beginning to the end of the paper, although there is evidence of some ramping within multi-part questions.
- In mathematics, the tests for 14 year-olds tests are tiered since no test is allowed to address more than three levels of difficulty in National Curriculum terms. This gives four pairs of test papers, Levels 3-5, Levels 4-6, Levels 5-7 and Levels 6-8. Each paper is 'ramped', beginning with easier questions and ending with harder ones, and contains a wide range of question types, which often extend over two or three pages. In 1995 and 1996, a calculator could be used in the test papers at each tier, although there were some questions for which it was specified that a calculator should not be used.

– In science, the tests for 14 year-olds are also tiered, and cover Levels 3-6 and Levels 5-7. All the questions in the common levels (Levels 5 and 6) appear in an identical form in both tiers, unlike some of the mathematics questions. The papers are intended to be divided into two sections, for example Levels 3 and 4 and Levels 5 and 6 in the Level 3-6 tier, so there is some ramping from the beginning to the end of each test paper. Again, the questions are varied in type and often extend over more than one page.

6.   THE ANALYSES

Four aspects of the work will be discussed in this chapter, in relation both to the mathematics and science outcomes.
1. A description of the new categories in terms of both content and skills.
2. A descriptive presentation of the results of applying these categorisations to the test questions (TIMSS and National Curriculum) and the conclusions to be drawn.
3. Analyses of performance data in relation to the new categorisations, both content and skills and the conclusions to be drawn.
4. Further analyses that may help to throw light on the explanation of the discrepancies.

7.   THE MAJOR FINDINGS IN MATHEMATICS

7.1   The categorisations

Questions were categorised into the following content and skill areas:[11.1]

Content areas	Skill areas
Whole numbers and number sense	*Knowledge*
Fractional number and proportionality	*Use of pre-learned procedures*
Measurement	*Problem solving*
Data representation, analysis and probability	*Mathematical reasoning*
Geometry	
Algebra	

[11.1] The TIMSS 'Content' categories were as follows:
  Population 1: Whole Numbers; Fractions and proportionality; Measurement, estimation and number sense; Data representation, analysis and probability; Geometry; and Patterns, relations and functions.
  Population 2: Fractions and number sense; Geometry; Algebra; Data representation, analysis and probability; Measurement; Proportionality.
  The TIMSS 'Performance Expectation' categories were as follows:
  Populations 1 and 2: Knowing; Performing routine procedures; Using complex procedures; Solving problems.

Figures 11.1, 11.2, 11.3 and 11.4 below give the details of the percentages in each category for the four sets of comparisons, namely the content areas and skill areas for the TIMSS tests and National Curriculum tests for both Population 1 and Population 2.

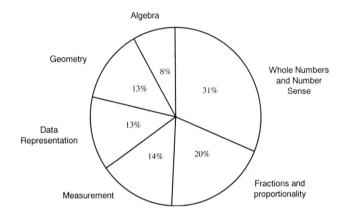

TIMSS Test questions Population 1 (9 year olds)

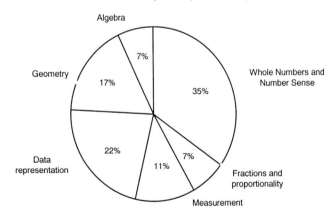

Key Stage 2
National CurriculumTest questions (1996 and 1997)

Figure 11.1. Pie charts showing the percentage of questions in each mathematics content area in the TIMSS tests (Population 1) and the National Curriculum tests for 11 year-olds

TIMSS Test questions Population 1 (9 year olds)

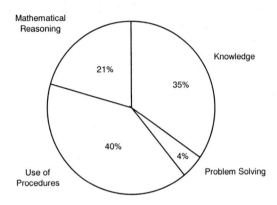

Key Stage 2
National CurriculumTest questions (1996 and 1997)

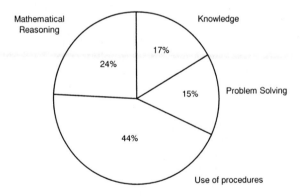

Figure 11.2. Pie charts showing the percentage of questions in each mathematics skill area in the TIMSS tests (Population 1) and the National Curriculum tests for 11 year-olds

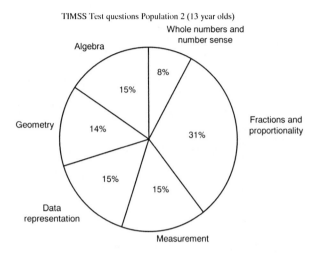

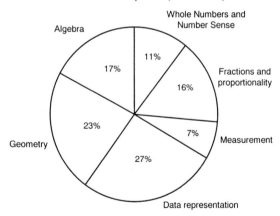

Figure 11.3. Pie charts showing the percentages of questions in each mathematics content area in the TIMSS tests (Population 2) and the National Curriculum tests for 14 year-olds

TIMSS Test questions Population 2 (13 year olds)

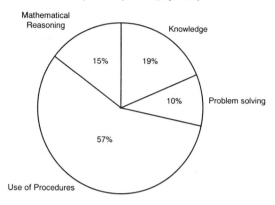

Key Stage 3 National CurriculumTest questions (1995 and 1996)

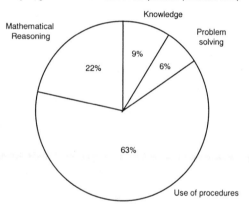

Figure 11.4. Pie charts showing the percentages of questions in each mathematics skill area in the TIMSS tests (Population 2) and the National Curriculum tests for 14 year-olds

The results of applying the categorisation system showed the following major points and differences in emphasis in the questions in the TIMSS tests and in the National Curriculum tests.

In relation to the content being assessed, comparison of the test questions in TIMSS Populations 1 and 2 with the relevant sets of National Curriculum tests revealed:

a) less emphasis on *Whole numbers and Number sense* and, in particular, on *Fractional number and proportionality* in the National Curriculum tests and;

b) more emphasis on *Data representation, Analysis and Probability* in the National Curriculum tests, and;

c) for the older age group (Population 2) more emphasis on *Geometry* in the National Curriculum tests.

In relation to the skills being assessed, there is less emphasis on knowledge in all the National Curriculum tests and more emphasis on problem solving (Key Stage 2) and mathematical reasoning and the use of procedures (Key Stage 3).

7.2   Pupil performance in the two kinds of tests in mathematics

These new categorisations of the content and skill areas of the two sets of question, open up the possibility of comparing performance on the various groupings of items using the TIMSS international average scores for the relevant age-groups and National Curriculum test performance data. For this kind of analysis, however, the most appropriate comparisons are for Population 2 (see earlier) and even then, the process is not without its problems since the two kinds of tests are different in their structure and purpose and there are no items in common between them. Preliminary analyses were carried out for Population 2 and the findings can be summarised in the following points.

– In the TIMSS tests for Population 2, the poorest performance (in relation to the international average performance) was in *Fractional number and Proportionality* and to a lesser extent in *Algebra* and *Whole numbers and number sense*.

– In the National Curriculum tests (14 year-olds) which were used for comparison purposes, English pupils' performance in *Whole numbers and number sense* was consistently higher than in other content areas, which may seem at face value to contradict the findings from the TIMSS tests which suggest this as an area of comparative weakness for English pupils. This may in fact be an indication that there is a range of factors to be taken into account when comparing performance, including the difficulty level and the type of questions. If children are performing less well than their international peers in an area in which they generally perform well on the National Curriculum tests, this is more likely due to the inherent nature of the questions than the 'ability' of the pupils.

In order to investigate the relationship between performance on the items in the two kinds of tests, a small-scale 'comparability' study was set up in which a sample of pupils worked with specially constructed test booklets, one for mathematics and one for science, containing items selected as being representative of the content balance and range of difficulty of the two tests. The results on this small sample showed that the National Curriculum tests contained a wider range of difficulty than the TIMSS tests but that the TIMSS tests (mathematics elements) for Population 2 seemed to be systematically more difficult in absolute terms than the broadly equivalent National Curriculum tests.

The explanation of the comparatively poor mathematics performance of the English pupils in the TIMSS tests is therefore a complex one, when seen in relation to the characteristics and outcomes of the national tests in mathematics for the relevant age groups. When both sets of questions (TIMSS and National Curriculum) were categorised according to the same set of criteria, important differences were revealed in both the content and skill areas of the questions. The TIMSS questions contained more questions on *Number* and on *Fractions and proportionality*, but there was more emphasis on *Data handling* in the National Curriculum tests. In terms of skill areas, the National Curriculum questions emphasised *Problem solving* (11 year-olds) and *Mathematical reasoning* and *Use of procedures* (14 year-olds). In so far as the tests embody the curriculum priorities and teaching emphases in schools and classrooms, these discrepancies could begin to explain the poor mathematics performance in relation to other countries in the TIMSS results.

8.   SUMMARY OF THE MAJOR FINDINGS IN SCIENCE

A parallel set of analyses was carried out for the science questions and these will be reported in the same way as for mathematics.

8.1   The categorisations in science

The new categorisations of the content and skill areas in science were as follows[11.2].

[11.2] The 'Content' categories for the TIMSS tests were as follows:
   Population 1: Earth science; Life science; Physical science; Science and the environment.
   Population 2: Earth science; Life science; Physical science; Chemistry; Science and environment (and other topics).
   The 'Performance Expectation' categories of the TIMSS tests, Populations 1 and 2 were as follows:
   Understanding; Theorising, analysing and solving problems; Using tools, routine procedures and science processes; Investigating the natural world; Communicating.

Content areas	**Skill areas**
Procedural knowledge	*Knowledge and understanding*
Biology	*Use of knowledge and understanding*
Chemistry	*Investigating*
Physics	*Drawing logical conclusions*
Uncategorised	

Note: The classification 'uncategorised' contains items, all from the TIMSS tests that would not fit this analytic framework. However, it should not be interpreted as a 'sink' category. Although the scientific content of these items was difficult to define (e.g. 'Write down one example of how computers help people to do their work'), the majority of the items would be seen as 'geographical' in the context of the National Curriculum. It would not be appropriate to classify these questions as involving science, but since there were significant numbers of them, they may have had an important effect on the outcomes. They therefore had to be included.

8.2   The test content and skills (TIMSS and National Curriculum tests compared)

The results of applying the categorisation system generated the following points and differences in emphasis in the questions in the TIMSS tests and in the National Curriculum tests in science.

– There were sizeable differences in the frequency of representation of the content and skill areas between TIMSS Population 1 and National Curriculum Tests for 11 year-olds. In particular, the emphasis on Biology in the TIMSS Population 1 questions contrasts with the more even balance of content areas in the National Curriculum tests. The proportion of TIMSS Population 1 items which sought to test Knowledge and Understanding was much higher than in the National Curriculum tests.

– In both the TIMSS Population 1 (9 year-olds) and KS2 tests, the content area of Chemistry was less well represented than either Biology or Physics.

– The differences were much less in evidence with TIMSS Population 2 and National Curriculum tests for 14 year-olds. With these older pupils, there was a relatively even spread of items between content areas in both tests and a similar balance in the skills tested. However, the substantial fraction of TIMSS Population 2 items which were Uncategorised is noteworthy and represents a significant imbalance in the distribution of items between content areas.

The data on which these comments are made are given in Figures 11.5, 11.6, 11.7 and 11.8 below.

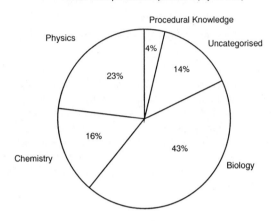

TIMSS Test questions Population 1 (9-year-olds)

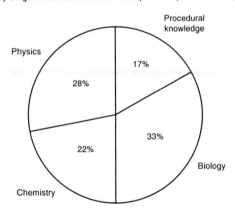

Key Stage 2 National Curriculum Test questions (1996 and 1997)

Figure 11.5. Pie charts showing the percentage of questions in each science content area in the TIMSS tests (Population 1) and the National Curriculum tests for 11 year-olds

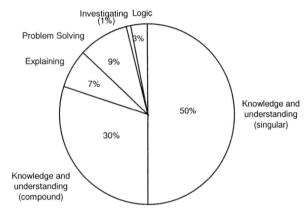

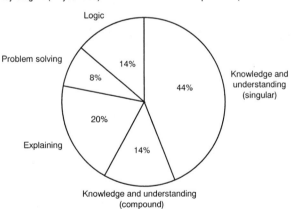

Figure 11.6. Pie charts showing the percentage of questions in each science skill area in the
TIMSS tests (Population 1) and the National Curriculum tests for 11 year-olds

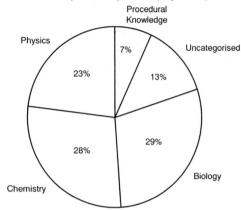

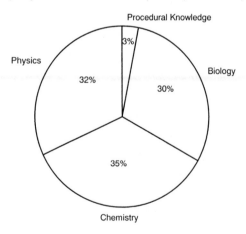

Figure 11.7. Pie charts showing the percentage of questions in each science content area in the TIMSS tests (Population 2) and the National Curriculum tests for 14 year-olds

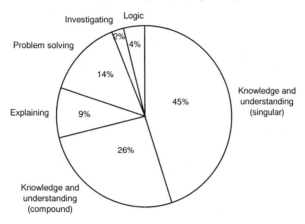

TIMSS Test questions Population 2 (13-year-olds)

Investigating   Logic

Problem solving

2%   4%

14%

Knowledge and understanding (singular)

45%

Explaining   9%

26%

Knowledge and understanding (compound)

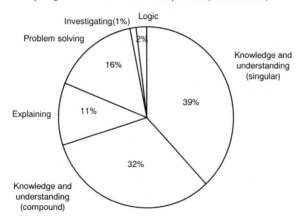

Key Stage 3 National Curriculum Test questions (1995 and 1996)

Investigating(1%)   Logic

Problem solving

2%

16%

Knowledge and understanding (singular)

39%

Explaining   11%

32%

Knowledge and understanding (compound)

Figure 11.8. Pie charts showing the percentage of questions in each science skill area in the TIMSS tests (Population 2) and the National Curriculum tests for 14 year-olds

8.3   Pupil performance in the two kinds of tests in science

An exactly parallel set of caveats must be entered here for the science as for the mathematics (see earlier). The most appropriate performance comparisons were those involving TIMSS Population 2 and the National

Curriculum tests for 14 year-olds but there were similar difficulties as with the mathematics. However, the initial investigation revealed some useful preliminary points which can be summarised as follows.

– In the TIMSS tests for Population 2, the English pupils outperformed the international sample (in relation to the international average) in every content area. However, with this age group, Biology, Physics and Uncategorised items elicited the best (and roughly equal) levels of performance, while items categorised under Chemistry and Procedural Knowledge produced slightly lower levels of performance.

– For Population 2 also, pupils in England out-performed the international average in each of the skill areas used in this study, with performance being strongest in the categories of Knowledge and Understanding, Explaining and Drawing logical conclusions.

– Allowing for the different numbers of items involved, the TIMSS Population 2 tests displayed broadly similar ranges of difficulty in each skill area. This was in marked contrast to the profile of the National Curriculum tests for 1996 (14 year-olds), which contained mostly poorly performed items within the skill category of Explaining.

As with the mathematics, therefore, the explanation of performance in the TIMSS tests is not straightforward.

In Population 2, English pupils achieved their best (if only by a short way) performance on Biology items, which was again the content area containing the most questions. The skill area of Knowledge and understanding again predominated in Population 2, although this did not elicit the best performance, a position given to Explaining. The data from the National Curriculum tests for 14 year-olds show that English pupils consistently perform well on Biology items, although this was not the case in the skill area of Explaining.

Since the TIMSS test emphasised areas of the curriculum which English pupils consistently perform well, this may go some way to explaining their better relative performance in the TIMSS tests generally.

9.   FURTHER ANALYSES THAT MAY THROW LIGHT ON THE DISCREPANCIES IN THE SCORING OUTCOMES

The final pieces of the jigsaw in trying to explain the TIMSS outcomes in the context of the National Curriculum and its testing are those of the types of questions found in the TIMSS and National Curriculum tests and the overall structure of the TIMSS test booklets. It has already been suggested earlier in this chapter that the majority of questions in the TIMSS tests were of the multiple-choice variety, a test format very unfamiliar to most English pupils. The test booklets also contained both mathematics and science questions, juxtaposed in different ways in the different pairs of test booklets. Again, this would be very unfamiliar to English pupils, but in this case perhaps to many pupils in other countries too. Each of these points is now considered in more detail.

9.1   Types of questions in the tests

If only one subject were being considered here, say mathematics, then it might be tempting to offer at least a partial explanation of the relatively poor performance of the English sample in terms of unfamiliarity with multiple-choice formats. However, this is not appropriate when both subjects are taken together. The design of the TIMSS survey encourages us to see rather different issues and to pose rather different questions of our own national curricula, pedagogic styles and assessments approaches. For instance, in the case of the mathematics and science outcomes in the TIMSS tests for the English sample, it cannot, in principle, be argued that the multiple-choice format is a problem for English pupils, when it seems to have less effect on their performance in one of the two subjects. In fact, however, the story is more complex than this.

More detailed investigation revealed that, in mathematics, approximately 80 per cent of all the questions were multiple-choice in format (77 per cent for Population 1 and 83 per cent for Population 2) with therefore only a small percentage of the question types most familiar to the English sample. The performance of the English sample on all question types follows broadly the same pattern as the International Average (in both age groups) except for extended answer questions where the performance of the English sample was better. However, the massive preponderance of multiple-choice type questions meant that any positive performance by the English samples on other question types would be considerably outweighed by the relative under-performance (in relation to the International average) on multiple-choice questions.

In science, multiple-choice questions also predominated with approximately 75 per cent of questions being of the multiple-choice variety (74 per cent for Population 1 and 75 per cent for Population 2). In science, however, the English samples out-performed many other countries and this performance is necessarily reflected in all question types, although the superior performance is less in comparison to performance on the 'open-response' questions.

9.2   The structure of the test booklets

In terms of the structure of the TIMSS test booklets, several points can be made, one of which also links to the question format analyses.   When identical questions appeared in different places in different tests very different responses were elicited and in fact, the lower performance on some of the open-response type questions (short and extended answer) may partly be explained by the location of many of these questions at the end of the test booklets.   In addition, however, some further issues should be discussed.

First, the structure of each booklet was very complex.   Not only were there questions in both mathematics and science, but these covered different content and skill areas in rapid succession.   For all pupils, in all participating countries, this could have been challenging.

Secondly, the questions in the TIMSS tests were not overtly 'ramped' in difficulty from beginning to end but this is a structure clearly evident in the National Curriculum mathematics tests for both 11 and 14 year-olds.   In the National Curriculum science tests, on the other hand, there was little evidence of a 'ramped' structure from beginning to end.   As such, the mathematics questions would be in a form much less familiar to English pupils, unlike the science.   To illustrate this point, Figures 11.9 and 11.10 below show the sequence of question difficulty from beginning to end of a typical pair of TIMSS test booklets and the National Curriculum tests for 11 year-olds in mathematics.

Booklets 5A and 5B

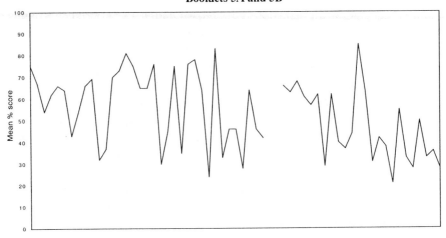

Figure 11.9. Performance on the questions in the order they appeared in a TIMSS
Population 1 pair of test booklets

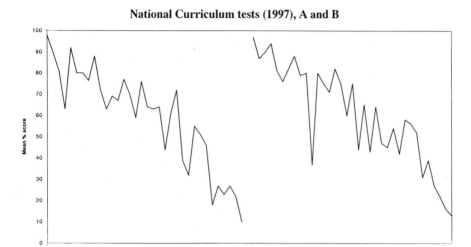

National Curriculum tests (1997), A and B

Figure 11.10. Performance on questions in the order they appear on National Curriculum
Key Stage 2 1997 Mathematics tests

9.3 The readability of the questions

Finally, could an explanation lie in the readability of the questions in mathematics and science that would help to account for the differences? There is much debate about the measurement of the reading difficulty of test questions in mathematics and to a lesser extent in science. Most traditional readability formulae are not readily applicable to mathematics text and particularly to the text of test questions. This is due to the use of specialised vocabulary, tables and graphs, labels and the specialist phrasing often used.

Table 11.5. The readability scores of the group of items in each content area in mathematics for the TIMSS Populations 1 and 2) and National Curriculum (Key Stages 2 and 3) test questions

	Whole numbers and number sense	Fractional number and proportionality	Measurement	Data representation and probability	Geometry	Patterns and relations (Algebra)
TIMSS 1	32.75	26.50	31.81	37.02	29.34	32.49
	Tok 1316	*Tok 472*	*Tok 491*	*Tok 800*	*Tok 384*	*Tok 379*
TIMSS 2	25.01	22.84	31.51	34.34	29.17	24.73
	Tok 622	*Tok 1811*	*Tok 799*	*Tok 1600*	*Tok 961*	*Tok 995*
Key Stage 2 11 year-olds	36.86	31.88	29.48	36.08	28.75	32.15
	Tok 554	*Tok 167*	*Tok 296*	*Tok 525*	*Tok 329*	*Tok 147*
Key Stage 3 14 year-olds	37.69	26.23	31.61	31.45	32.46	25.69
	Tok 613	*Tok 977*	*Tok 302*	*Tok 2404*	*Tok 1375*	*Tok 1290*

Note: Tok = tokens, i.e. the number of word/symbols used for the calculation.
Where this is lower than 400, the data are less reliable.

This also applies to scientific text in many ways. The only existing formula that seems appropriate for these purposes in tests is the Kane Formula 2, although even this is not without its problems (see Shorrocks-Taylor and Hargreaves, 1999). The results of applying these calculations to a sample of both the mathematics and science questions are shown in Tables 11.5 above and 11.6 below (the higher the value in the cells, the easier to read).

Table 11.6. The readability scores of the group of items in each content area in science for the TIMSS Populations 1 and 2) and National Curriculum (Key Stages 2 and 3) test questions

	Procedural Knowledge	Uncategorised	Biology	Chemistry	Physics
TIMSS 1.	33.88	30.44	30.34	30.68	30.56
	Tok 234	*Tok 415*	*Tok 1252*	*Tok 494*	*Tok 699*
TIMSS 2	26.64	26.60	24.89	27.91	30.21
	Tok 652	*Tok 695*	*Tok 1483*	*Tok 1443*	*Tok 1409*
Key Stage 2	31.36		29.15	34.00	32.06
11 year-olds	*Tok 523*	None	*Tok 332*	*Tok 572*	*Tok 639*
Key Stage 3	31.13		26.84	24.60	29.92
14 year-olds	*Tok 361*	None	*Tok 1685*	*Tok 2075*	*Tok 2308*

Note: Tok = tokens, i.e. the number of word/symbols used for the calculation.
Where this is lower than 400, the data are less reliable.

The broad conclusion to be drawn from the information in these Tables is the great variation in the mean readability scores of the questions in the content areas in both science and mathematics. In more detail, the major points can be summarised as follows.

a) For some content areas in both mathematics and science, the readability scores on the three different sets of questions were closely grouped together [*Measurement* (mathematics), *Physics* (science) and to a lesser extent '*Geometry*' (mathematics)].

b) For other content areas in both subjects, the scores were widely spread [*Whole numbers and number sense* (mathematics), *Fractional number and proportionality* (mathematics), *Patterns and relations* (mathematics), *Procedural knowledge* (science) and *Chemistry* (science)].

c) In mathematics, the most difficult questions to read occur in *Whole numbers and number sense, Fractional number and proportionality* and *Patterns and relations (Algebra)* and these occur in the TIMSS Population 2 questions.

d) In science, the most difficult questions to read occur in *Biology* and *Chemistry* but, here, both the TIMSS Population 2 questions and the Key Stage 3 questions are involved.

e) In science, there is a pattern to be detected in that the hardest questions in terms of reading demand, across the content areas, occur in the TIMSS Population 2 tests and in the Key Stage 3 tests, both of which are aimed at the older age groups of pupils: the tests aimed at the younger pupils (TIMSS Population 1 and the Key Stage 2 tests) are systematically easier to read.

f) In mathematics on the other hand, such a pattern is less easy to detect: in the main it is the TIMSS Population 1 tests and those at Key Stage 2 which prove easier to read in the different content areas, but there are exceptions to this *in Whole numbers and number sense* and in *Geometry*.

g) In general, the mathematics questions seem to be slightly easier to read than the science questions, but this is a very small difference.

Although interesting in their own right, these results on readability throw only limited light on the matter of accounting for the performance of the English samples. The TIMSS questions in both mathematics and science were varied in their ease/difficulty of reading, and there is even some suggestion that the mathematics questions were marginally easier to read.

10. DISCUSSION

This study began as a commissioned investigation into the TIMSS results published in England in 1996 (for Population 2) and 1997 (for Population 1). The public perception of the reported findings was that both our 9 year-olds and 13 year-olds had performed poorly in mathematics but better in science. This differential subject performance was given considerable emphasis in the press.

What is clear from the published TIMSS data, however, is that the results should not be considered in this over-simplistic fashion. The tests were not devised to yield similar distributions and mean scores in the two subjects and the rank orderings of the participating countries should not therefore be interpreted in this absolute kind of way. Inevitably, rank orderings are the product of a complex range of performances within and between all the samples of students and often just a small number of marks separate countries in the listings. Only a much more detailed analysis than those published so far would allow such statements of differential performance in the two subjects to be made with any rigour.

The main emphasis of this study subsequently became the comparison of the TIMSS test outcomes with those of the National Curriculum testing in mathematics and science, considered independently rather than in relation to each other. It was hoped that such comparisons would yield insights that would both help to explain the English performance and provide pointers towards the further development of the National Curriculum for the year

2000. Five research questions were posed earlier (p.235) which can now be addressed. The answers to the questions have proved by no means straight-forward: in fact we have probably raised more questions than we have answered. This is not necessarily unhelpful, since over-simplistic responses would be unhelpful on all counts. However, some tentative conclusions can be reached.

Research questions 1 and 3 focused on the mathematical and scientific content and skill areas of the questions in the two kinds of tests. In both subjects, the categorisations (different from those used in the TIMSS study) revealed systematic differences, reflecting the different priorities in the English National Curriculum. In mathematics, there was less emphasis on number, fractions and proportionality in the National Curriculum and more emphasis on data handling, problem-solving and mathematical reasoning. In science, the National Curriculum tests showed a better balance across the major content areas than the TIMSS tests and less emphasis on the straightforward recall of knowledge.

These differences are important, but their implications can only be revealed fully if performance in the different content and skill areas is incorporated into the equation, hence the focus of research questions 2 and 4. Given two very different kinds of testing regime, each with different purposes, direct performance comparisons are not easy in either mathematics or science, and the reasons for this are discussed earlier.

Research question 5, looking at wider issues, proved to be quite a rich seam to mine. The analyses presented earlier focused on three aspects, namely the effects of question types, the structure of the tests and the readability of the questions. The TIMSS tests represented a very particular approach to testing: they were predominantly of one question type – multiple-choice – and were organised into booklets in demanding ways that were perhaps unfamiliar to many pupils. On the other hand, the National Curriculum tests in England and Wales had (and have) a very different character: the questions were more varied in format, more open-ended and employed a greater variety of presentation and response modes than the TIMSS tests. Even in 1995, 14 year-old pupils were fairly familiar with this approach to testing and 11 year-olds had some experience too.

This appears to offer some explanation of the performance of the English samples of pupils, until the question of the differences in mathematics and science performance is raised. The only tentative suggestion offered here is that English pupils' experience of testing in science was more closely related to the TIMSS approach, in not being 'ramped' and in containing more questions of a broadly multiple-choice character.

The investigation of the readability of the questions, whilst interesting and important in its own right, does not really provide an answer to the basic question posed here. Measuring the ease or difficulty of reading test questions is an area of research still in its infancy, yet it is one that merits much more attention. It is part of the wider assessment research issue of needing to understand a great deal more about exactly how pupils, especially younger ones, approach test questions: how and in what kind of sequence do they read and interpret a question and how do they marshal their knowledge and thoughts to be able to provide a response? Only much more research information on these points will help us to devise questions in a more focused and relevant way.

However, whatever the interpretation and explanation, the TIMSS (1995) results for the English samples of pupils stand. It will be interesting and informative to see if the results from the re-run of TIMSS in 1999 produce a similar pattern of scoring in England, since this in itself may help to answer some of these questions. The forthcoming OECD triennial surveys may also give rise to other insights. Perhaps what is really needed at an international level is a comparative study of the comparative studies, looking for points of comparability and non-comparability between them. If we were all clear what we really need to know, for instance about mathematics attainments, then it should be possible to seek out or generate more relevant data to find the answers.

Our final comment must therefore be that the TIMSS results have provided a point of discussion and challenge within the English education system at a time when the National Curriculum and pedagogic practice in schools is under review. The fact that there is a well-established and extensive national assessment system here in England means that comparing outcomes proves both interesting and useful, with, it is hoped, a two-way learning process taking place with the researchers in the international comparative studies.

ACKNOWLEDGEMENTS

We wish to acknowledge the contribution to the work on which this chapter is based of our co-researchers, Melanie Hargreaves, Peter Laws and Nick Nelson.

REFERENCES

Harris, S., Keys, W. and Fernandes, C. (1997) *Third International Mathematics and Science Study: Second National Report* (Slough: NFER)

Keys, W., Harris, S. and Fernandes, C. (1996) *Third International Mathematics and Science Study: First National Report, Part 1* (Slough: NFER)

Keys, W (1999) What can mathematics educators in England learn from TIMSS? *Educational Research and Evaluation*, 5, pp. 195-213

Shorrocks-Taylor, D and Hargreaves, M (1999) Making it clear: a review of language issues in testing, with special reference to the National Curriculum mathematics tests at Key Stage 2. *Educational Research,* Vol 41, No 2, pp. 123-136

Chapter 12

Messages for Mathematics Education from TIMSS in Australia

Jan Lokan
Australian Council for Educational Research, Australia

Australia has a long history of participation in studies carried out under the auspices of the International Association for the Evaluation of Educational Achievement (IEA), particularly in mathematics and science. It was one of only twelve countries involved in the inaugural mathematics study in 1964 (although one of the six main states did not participate and only government schools were included in the samples); a replication of this study was carried out within Australia in 1978; and the country participated fully in the first and second international science studies. The international centre for the second science study was based at the Australian Council for Educational Research (ACER), directed by Dr Malcolm Rosier. It would thus have been surprising if Australia had not taken part in TIMSS, though participation on a national basis required agreement from nine governments (the federal government, six state governments and two territory governments).

1. HOW WE PERFORMED: THE FIRST AND SECOND STUDIES

In 1964, the mean mathematics score for Australian grade 8 students was higher than that for students in the same grade in the USA and Sweden, but was lower than that for same-grade students in Japan and several European countries, including England and Scotland (Spearritt, 1987). Already, this result highlights the difficulties of making international comparisons, as there are other factors that need to be considered in interpreting the findings.

D. Shorrocks-Taylor and E.W. Jenkins (eds.), Learning from Others, 259–277.

For example, the Australian students were almost a year younger than the students at the same grade level in most other countries. In addition, the participants were representative of the government education sector only. A fully representative sample would have been expected to achieve at a higher level, as a fifth to a third of the secondary population is educated in non-government schools and, on average, students from the non-government sector traditionally perform better than students from the government sector.

In the later IEA studies of mathematics and/or science, the Australian samples were more representative. In the first science study, carried out in 1970-71, Australian 14 year-old students were ranked third of 14 developed countries. In the second science study, carried out in 1983-84, Australia's relative achievement was similar, placing us in a group of seven countries which ranked third, behind Hungary as clearly the highest achieving country and Japan and the Netherlands, which formed the second highest achieving group. Keeves (1992) placed the results of the ten countries that participated in both the first and second science studies on a common scale and found that Australia's results remained constant, in absolute terms. By contrast, the USA's results declined and the results of the other eight countries improved.

Australia did not participate in the 1980-81 international mathematics study, in which newly developed tests were used. Instead, the 1964 tests were administered, this time to a more representative sample, in 1978. This enabled changes from 1964 to 1978 to be assessed in the government sector in the five states taking part in both studies, and an estimate of achievement to be determined for the country as a whole. A slight but significant overall decline was found except for Western Australia, with geometry achievement declining more than algebra or arithmetic achievement (Rosier, 1980).

In the context of the possibly declining achievement in mathematics and static achievement in science identified in the first and second studies, results from TIMSS were eagerly awaited by Australian educators. Many changes in curricula had been implemented since the second studies. Had these led to improved performance? The substantial curriculum revision work at national and state levels in the early 1990s had not reached the schools on a wide scale when the TIMSS data were collected, but the TIMSS Expert Questionnaires identified significant changes, mainly in teaching approaches, over the previous 15 years (Lokan, Ford & Greenwood, 1996). Essentially, the changes were to place emphasis on the relevance and applications of mathematics in everyday life, with instruction based on problem solving and investigative approaches instead of on formal algorithms. In recent years, the use of computers and calculators to aid in the development of concepts has been encouraged at both primary and secondary level in all Australian education systems.

2.   SCOPE OF TIMSS IN AUSTRALIA

Australia participated in all aspects of TIMSS at each of the three populations, including the Performance Assessment (PA) component. In one sense, the PA component was crucial to the success of the study, as it assisted national project staff to convince the teacher unions that TIMSS would go beyond earlier studies and be worthy of their support. Australian schools have autonomy in deciding whether to participate in research, and opposition from unions is known to have influenced response rates in other national and state-based studies (McGaw, Long, Morgan & Rosier, 1989).

2.1   Student samples

Policies on the age at which children are eligible to start school vary among the Australian states and territories (hereafter referred to as states), which meant that the two grade levels containing the most 9 year-olds and the most 13 year-olds also varied. For Population 1, grades 3 and 4 were sampled in half the states and grades 4 and 5 in the others. For Population 2, likewise, grades 7 and 8 and grades 8 and 9 were the pairs of adjacent grades sampled. For Population 3, students in Year 12 (about 70 per cent of the cohort) were sampled from all states.

Altogether, over 11,000 and almost 14,000 students were tested at Population 1 and Population 2, respectively. Two classes per school were tested at grade 4 and grade 8, the modal years for 9 year-olds and 13 year-olds, irrespective of whether these were the upper or lower grade for their population. Two classes per school were also tested at grade 5 and grade 9, but only one class per school was tested at grade 3 and grade 7. In determining the results, statistical weighting adjusted for the differing representation of grade levels. About 3,200 students, sampled individually within the selected schools, were tested at Population 3.

The testing was conducted in October and November 1994 (Populations 1 and 2) and August to October 1995 (Population 3).

2.2   School and teacher samples

The Population 1 students who participated in TIMSS came from 542 classes in 179 schools, the Population 2 student participants came from 587 mathematics classes in 180 schools and the Population 3 students came from 90 schools. All education sectors from all parts of Australia were represented in the samples.

2.3   Adequacy of samples

The weighted response rates of selected Population 1 and Population 2 schools and students did not quite reach the internationally desired level, falling short by about five per cent. From locally available information, the responding and non-responding schools were compared on a range of socio-economic indices and on data from state-wide testing programs. The conclusion from these comparisons was that the Australian TIMSS primary and lower secondary level results are very unlikely to be biased because of the slightly lower than required response rates. At senior secondary level this claim cannot be made, as there was no possibility of conducting similar comparisons for the Population 3 school sample.

3.   HOW WE PERFORMED: TIMSS

On the whole, Australian students performed relatively well, equal to or better than their peers in other English speaking countries and in many European and other countries. However, at Populations 1 and 2, the students from the neighbouring Asian countries of Singapore, Korea, Japan and Hong Kong generally outperformed our students by quite a large margin. These countries did not participate at Population 3.

In mathematics:
– Six of 25 and four of 23 other countries performed better than Australia at the international fourth and third grades, respectively (Singapore, Korea, Japan and Hong Kong at both grades, plus The Netherlands and the Czech Republic);
– Eight of 40 and seven of 38 other countries performed better than Australia in the international eighth and seventh grades, respectively (Singapore, Korea, Japan, Hong Kong, Flemish Belgium and the Czech Republic at both grades, plus the Slovak Republic and Switzerland at eighth grade and The Netherlands at seventh grade);
– Australia was one of ten countries all achieving similarly at the final secondary level in advanced mathematics, with no country performing better;
– Only one country (The Netherlands) performed better than Australia at the final secondary level in 'mathematics literacy'.

In science:
– Two of 25 and only one of 23 other countries performed better than Australia at the international fourth and third grades, respectively (Korea at both grades, plus Japan);

- Four of 40 and seven of 38 other countries performed better than Australia in the international eighth and seventh grades, respectively (Singapore, the Czech Republic, Japan and Korea at both grades, plus Bulgaria, Slovenia and Flemish Belgium at seventh grade);
- Only two countries (Norway and Sweden) performed better than Australia at the final secondary level in physics;
- No country performed better than Australia at the final secondary level in 'science literacy'.

3.1   How were these results received?

Two weeks before the first TIMSS results were released, a Melbourne daily newspaper featured the following:

Crisis forecast in maths
A recent study of Australian children's maths. and science skills was likely to paint a gloomy picture, a mathematician warned yesterday ... "We're bracing ourselves for the bad news", she said. (*Herald Sun*, 6 November 1996, p. 2)

On or shortly after the release of the Population 2 results, reception was varied, ranging from 'Our pupils near top of the class in maths, science' and 'Australians in top 10 at science and maths' (referring to the total results); and 'WA students outsmart most' (in the national breakdown of results, Western Australia performed exceptionally well, at the same level as the top countries); to 'Australia lags in school survey'; 'Maths study shows cause for concern' (referring to the total results); and 'State flops in maths, science: Dearth of subject teachers revealed' (referring to one of the lowest achieving states, which nevertheless still performed at the international average).

Even though not given a prominent place in the newspaper, the most influential reaction came from the Commonwealth Minister for Schools. In an article headed 'Maths, science push', the Minister was quoted as saying: 'Australian students' maths and science standards needed to be raised if Australia was to keep pace with its Asian neighbours'. The Minister used the opportunity to announce that 'The States and the Commonwealth are now working closely together to establish national benchmarks in literacy and numeracy'. (*Sunday Mail*, 24 November 1996, p. 39)

Since that time the focus of comment and reform efforts in Australia has been firmly on mathematics rather than science, with large energies expended in establishing 'benchmarks' in numeracy for students in Years 3, 5, and 7. (In keeping with the relatively better performance in science than in mathematics, an interest in science education reform is only now beginning to emerge, three years after the first release of TIMSS results.) In the climate of the current concern for what local educators call 'numeracy', but which is assessed within school mathematics, the remainder of this chapter presents some messages for mathematics education in Australia that can be gleaned from the primary and lower secondary students' responses to the TIMSS tests.

4. AREAS OF STRENGTH AND WEAKNESS IN MATHEMATICS

4.1 Results by content area

At Population 2, Australian achievement was furthest above average in the mathematics area of 'data representation, analysis and probability' and closest to the international average in 'proportionality' and 'geometry'. The result for geometry was particularly interesting, but disappointing, for our mathematics educators because this was one of the two areas in which our students performed relatively well at Population 1 (the other was 'data representation, analysis and probability'). Data representation, analysis and probability all receive emphasis in current Australian mathematics curricula. The low emphasis on geometry identified for Australia in the curriculum and textbook analyses for Population 2 (Schmidt *et al.*, 1997) may have contributed to the differential performance in this area between the Population 1 and Population 2 samples. Geometry was also the topic done least well by the Australian advanced mathematics students at Population 3. The extent of differences between the Australian and international average performance at the international fourth and eighth grades is illustrated in Figure 13.1 below. To meet the age band criteria for TIMSS sampling, in Australia these were mixtures of Years 4 and 5 and Years 7 and 8, depending on the state.

4.2 Some diagnostics at item level

It is both more interesting and more useful, from the point of view of messages for mathematics teaching, to analyse the TIMSS results at item level.

A comparison with the international average per cent correct for each of the 157 TIMSS Population 2 mathematics items, taking a 5 per cent difference as likely to be significant, showed that the Australian upper and lower grade students performed above the international average on 66 and 67

Population 2

Average per cent correct

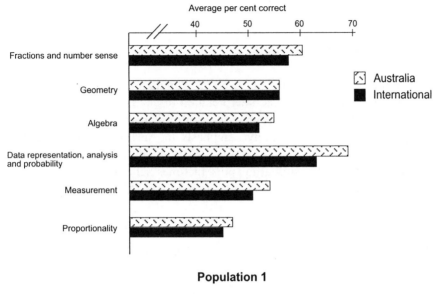

Population 1

Average per cent correct

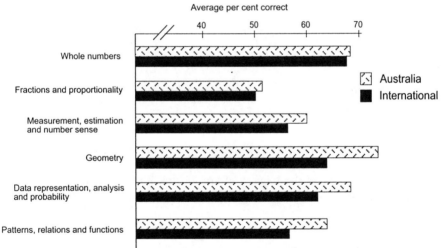

Figure 13.1. Australian and International Results in TIMSS Mathematics Content Areas

items, respectively. Australian students' performance was significantly lower than the international average on only 15 items at the upper grade and 18 at the lower grade. Usually similarly good or poor performance occurred on the same items at both grade levels.

A similar analysis for the 107 Population 1 mathematics items showed students' performance to be above the international average on 47 and 35 items at the upper and lower grades, respectively, and to be below the international average on only nine items at the upper grade and eleven at the lower grade.

4.1.1   Strengths, Population 2

The items done particularly well at Population 2 primarily involved conceptualisation of fractions, interpretation of charts and graphs, rounding, estimation, some algebra and most probability items. These areas, with the exception of algebra, all receive emphasis in Australian lower secondary level mathematics curricula.   On these kinds of items Australian performance was often similar to and sometimes better than the performance of students in the high achieving Asian countries. An example is shown in Figure 13.2.

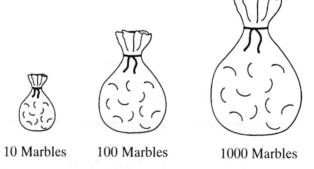

There is only one red marble in each of these bags.

10 Marbles   100 Marbles   1000 Marbles

Without looking in the bags, you are to pick a marble out of one of the bags. Which bag would give you the greatest chance of picking the red marble?

A. The bag with 10 marbles   C. The bag with 1000 marbles
B. The bag with 100   D. All bags would give the same
                                     chance.

Figure 13.2. Example of a TIMSS Item Done Well by Australian Students at both
Population2 and Population 1

The international average percentage correct on this item was 75, for Australia the percentage correct was 84, for the four Asian countries it was 86 and for England it was also 86.

4.1.2 Strengths, Population 1

Many geometry items were among those done best at Population 1. Australian students demonstrated good understanding of the properties of geometrical figures, in 3-D as well as in 2-D, and performed well on items involving concepts of symmetry or rotation, knowledge of sides and faces of cubes, how to read grids and understanding congruence in an elementary way. In other areas, they showed good understanding of number patterns, conceptualisation of fractions, reading information from graphs and elementary probability. The item illustrated in Figure 13.2 was also included in the Population 1 test, and the students achieved a similarly good result (59 per cent correct compared with 60 per cent in the four Asian countries and an international average of 51 per cent). An example of a Measurement item on which Australian students performed well is shown in Figure 13.3 and another probability item with similar good results is shown in Figure 13.4. Both of these items received high ratings for relevance to Australian curricula.

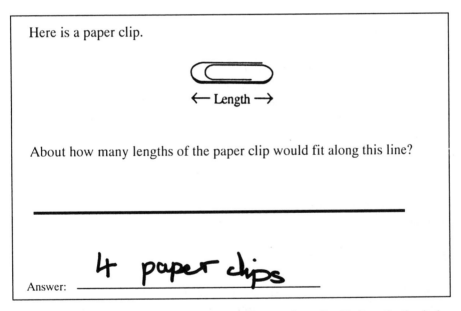

Figure 13.3. Example of an Item Judged High Relevant to Australian Mathematics Curricula

The international average percentage correct on this item was 48, for Australia and England the percentage correct was 58, and for the four Asian countries it was 59.

Samantha drops a stone onto each of these targets. In which target does the stone have the best chance of landing on a shaded space?

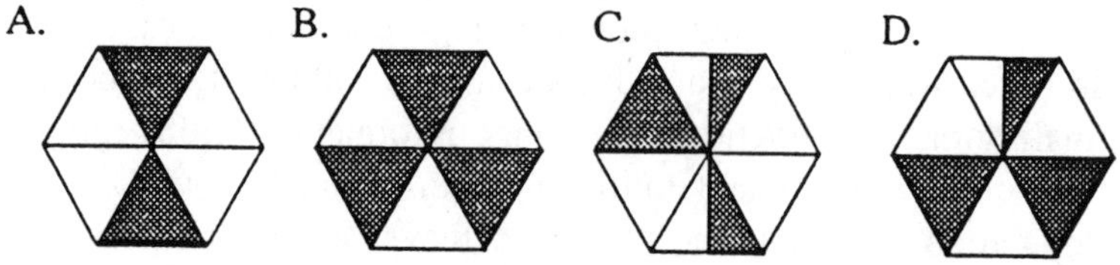

A.   B.   C.   D.

Figure 13.4. Example of a Population 1 Item done well in many countries, including Australia and England

The international average percentage correct and the percentage correct for Australia, England and the four Asian countries were all about 80.

4.1.3   Weaknesses, Population 2

The items with which Australian Population 2 students had most trouble were primarily multiplication and division of decimals and fractions, items requiring more than two fractions to be combined in some way, and items requiring more than one step to solve. Overall, Australia's performance in 'fractions and number sense' was a little above the international average; had the students been as successful on the more difficult operations with fractions and decimals, and with computations generally, as they were with conceptualisation and estimation, their performance would have been exceptionally good in this area. Some examples are provided in the detailed section on weaknesses below.

4.1.4   Weaknesses, Population 1

At Population 1, the students had most difficulty, relative to other countries, with addition and subtraction of four-digit whole numbers, subtraction of simple decimals, a word problem requiring multiplication of a three-digit whole number by nine, an item requiring interpretation of an inequality (this item was very well done in Asian and most European countries) and an item involving expressing an addition fact in terms of a multiplication fact. Again, some examples are provided in the following section.

5.   DETAILS OF THE WEAKNESSES IDENTIFIED

The single most prominent factor in our lower achievement in relation to the four Asian countries, the aspect of our achievement that had concerned the Commonwealth Education Minister so much, was Australian students' inability to perform computations accurately. On some routine computation items, the Australian students actually achieved the lowest result among the developed countries in TIMSS. (Typically, though, if Australian students performed poorly on a computation or problem solving item, students from the UK and New Zealand performed at an even lower level.) Abysmally poor performance was not uniform throughout the Australian states. While the level of performance on these kinds of items was poor in all states relative to the Asian countries, students from the Australian Capital Territory and South Australia (and sometimes also Western Australia) tended to perform rather better than students from the other Australian states.

Table 13.1 presents comparative results for the international fourth and eighth grades on some of the items where the discrepancy between Asian students' performance and Australian performance was greatest.

On a Population 2 item involving division of decimals ($0.004\overline{)24.56}$), which, for a student understanding the process would have meant no more than dividing by 4, only 23 per cent of Australian students were successful, compared with the international average of 44 per cent correct and an average for the Asian countries of 68 per cent correct. Only students from Colombia, Iran, Kuwait and Scotland scored at a significantly lower level than Australia on this item. On another item, which asked students to write 0.28 as a fraction in its lowest terms, Australian students (40 per cent correct) were able to perform above the international average (32 per cent correct), but were well below Hong Kong, Japan and Korea (about 70 per cent correct) and Singapore (90 per cent correct). Analysis of about 100 students' responses to this open-ended question showed that usually the students knew what they needed to do (although some clearly had no idea – see Example 3 in Figure 13.5 below), but simple computation let them down. Some examples of their responses are presented in Figure 13.5.

A wide variety of answers, several obviously resulting from problems with computation, was offered for this question. Responses with 25 in the denominator but 5, 6 or 8, or even 14, in the numerator were quite common. So were $^1/_3$ and $^1/_4$, $^4/_7$ and $^2/_8$ or $^3/_8$. There were some innovative 'Let's try something with the numbers we have been given' responses, including $^{.28}/_{10}$, $^{28}/_{10}$, $^{28}/_1$, $^1/_{28}$ and $^2/_{0.8}$.

Table 13.1. Results on TIMSS Items with Large Discrepancies Between Australian and Asian Students' Performance

Item	Item type*	Percentage correct					
		International average	4 Asian countries' average	Singapore	Australia	England	New Zealand
Population 2							
$\frac{3}{4} + \frac{8}{3} + \frac{11}{8} =$	MC	50	85	91	35	12	20
Divide: $0.004\,)\overline{24.56}$	MC	44	68	68	23	19	23
Write 0.28 as a fraction reduced to its lowest terms	MC	34	73	90	40	21	22
Multiply: $0.203 \times 0.56 =$	CR	49	64	66	25	8	21
Mr Lewis had $360. He spent $\frac{7}{9}$ of it. How much money did he have left?	CR	32	61	80	34	26	28
Find x if $10x-15 = 5x+20$	CR	45	78	79	34	22	19
Divide: $\frac{8}{35} \div \frac{4}{15}$	CR	43	79	87	25	8	9
Population 1							
$6000 - 2369 =$	MC	71	91	92	47	36	30
$6971 + 5291 =$	MC	84	94	95	76	61	69
Addition fact: $4 + 4 + 4 + 4 + 4 = 20$ Write this addition fact as a multiplication fact ____ × ____ = ____	CR	77	93	90	71	53	67
In which pair of numbers is the second number 100 more than the first number?	MC	49	71	61	44	34	33
25×18 is more than 24×18. How much more?	MC	45	69	73	40	37	28

* MC = multiple choice; CR = constructed response

Item: Write 0.28 as a fraction reduced to its lowest terms.

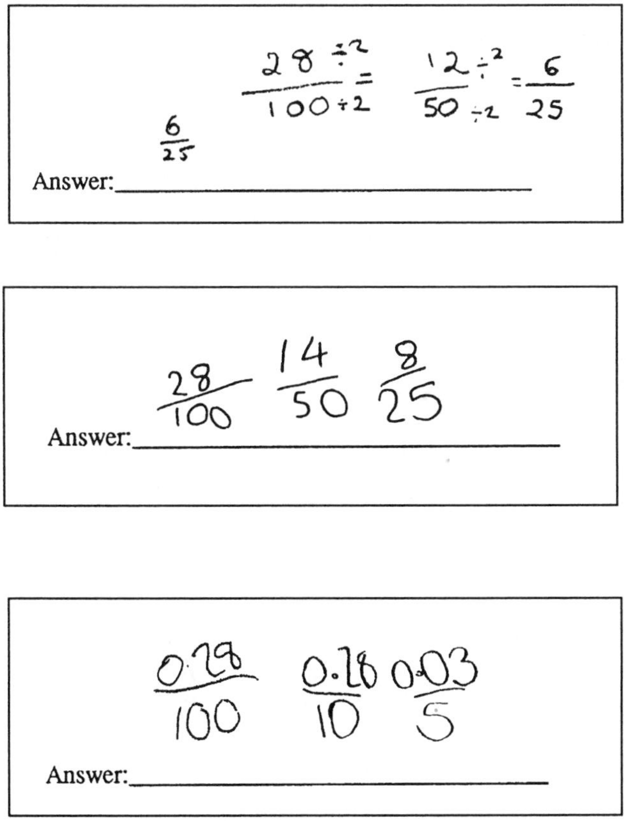

$$\frac{28 \div 2}{100 \div 2} = \frac{12 \div 2}{50 \div 2} = \frac{6}{25}$$

$$\frac{6}{25}$$

Answer:_____

$$\frac{28}{100} \quad \frac{14}{50} \quad \frac{8}{25}$$

Answer:_____

$$\frac{0.28}{100} \quad \frac{0.28}{10} \quad \frac{0.03}{5}$$

Answer:_____

Figure 13.5. Example of a Population 2 Item on which many Australian students performed poorly

Although Australian performance in algebra at Population 2 was quite good, there were some items where this went against this general trend. One was the 'solve for x' item, with 'x' on both sides of the equation, shown in Table 13.1. Another, which caused difficulty for about half the students, was the item shown in Figure 13.6. Given the problem with manipulating fractions that is evident for the item in Figure 13.5, it is not surprising that many students were not able to obtain the correct answer when substituting '2' for 'x' in the item in Figure 13.6.

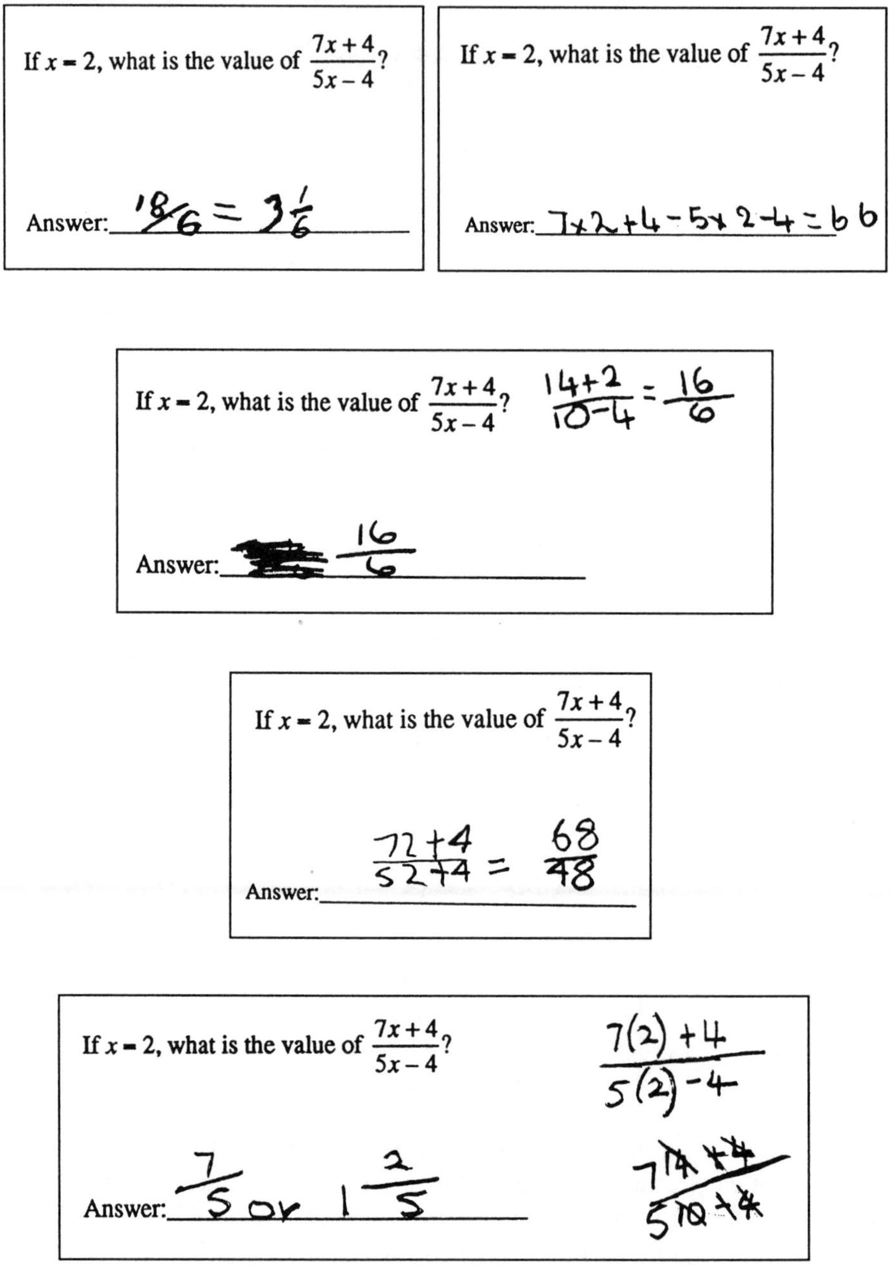

If $x = 2$, what is the value of $\dfrac{7x+4}{5x-4}$?

Answer: $\dfrac{18}{6} = 3\dfrac{1}{6}$

If $x = 2$, what is the value of $\dfrac{7x+4}{5x-4}$?

Answer: $7 \times 2 + 4 - 5 \times 2 - 4 = 6\,6$

If $x = 2$, what is the value of $\dfrac{7x+4}{5x-4}$?　$\dfrac{14+2}{10-4} = \dfrac{16}{6}$

Answer: $\dfrac{16}{6}$

If $x = 2$, what is the value of $\dfrac{7x+4}{5x-4}$?

Answer: $\dfrac{72+4}{52+4} = \dfrac{68}{48}$

If $x = 2$, what is the value of $\dfrac{7x+4}{5x-4}$?

Answer: $\dfrac{7}{5}$ or $1\dfrac{2}{5}$

$\dfrac{7(2)+4}{5(2)-4}$

$\dfrac{7\,14+4}{5\,10-4}$

Figure 13.6. Example of a Population 2 Item on which many more Australian students performed poorly

An amazing array of incorrect answers was found for this item in a sample of only about 100 test books. Answers included:

$2x$ (with working $7x - 5x$);

$12x$ (with working $7x + 5x$);

$13x$ (with no working);

$7x/5x$ (with working showing the '4's cancelled out);

$7/5$ (with working showing both the 'x's and the '4's cancelled out);

4, 6, 12, and 24 (each with no working);

124 (with working $72 + 4 - 52 - 4 = 124$);

126 (with working $78 + 58 = 126$);

$1^3/_8$ (with working $^{76}/_{48} = 1^3/_8$);

$^3/_2$ (with working $^{18}/_6 = ^3/_2$);

$^8/_3$ (with working $^{18}/_6 = ^8/_3$);

$^{76}/_{56}$, $^{76}/_{48}$, $^{18}/_{16}$ and $^{18}/_{14}$ (each with no working, but found several times);

$^{13}/_3$, $^9/_7$, $^{10}/_6$, $^{11}/_6$, $^{18}/_8$, $^{14}/_{20}$, and $^{18}/_{46}$ (each with no working, but found less often).

About half of these result from a failure to make the substitution correctly, thinking, for example, that $7x = 72$ when $x = 2$, which could arise from lack of exposure to algebra. The other half are most likely due to mistakes in simple computation, to the point of engendering despair that these students will ever become numerate citizens. The TIMSS Test Curriculum Matching Analysis, carried out by curriculum specialists in each state, confirmed that the intended curriculum in three of the eight Australian states and territories in 1994 did not include algebra beyond simple linear equations with the unknown, x, on one side. This should have been sufficient for the notation '$7x$' in this item to be understood, however.

Some Australian mathematics educators say that they are not concerned about the decrease in routine computational skills that was expected to happen when curricular emphasis on this aspect of mathematics was reduced a decade or so ago. After all, students are expected to use calculators these days, and calculator use is encouraged from first grade in some states. Others, especially mathematicians in universities and employers generally, lament the lack of skills that students bring with them to their jobs or to their education beyond school. It seems entirely reasonable to expect students in Grade 8 to be able to divide by 2 to reduce $^{14}/_{50}$ to $^7/_{25}$, or to be able to divide by 6 (or 2 and then 3) to reduce $^{18}/_6$ to 3.

It is claimed that Australian mathematics curricula now emphasise problem solving, conceptual understanding and reasoning, to replace the emphasis on arithmetical operations as curricula were revised in the 1980s. Yet Australian students did not perform particularly well on many of the problem solving items in TIMSS, usually being outscored substantially by students from Asian countries just as they were on computational items.

A detailed study of a sample of responses to open-ended problem solving items from about 300 students was undertaken. This revealed that students who made errors in solving verbal problems usually did understand the task they needed to do, and often managed to carry out one or even two steps correctly. Typically, they came unstuck once again with relatively simple computations, such as subtracting 280 from 360, as can be seen in the example presented in Figure 13.7. Only 34 per cent of the Australian students were able to work through this item to the correct answer. The item was difficult in all countries except Singapore, where 80 per cent of the students answered it correctly. Next highest was Japan, with 57 per cent correct.

Item: Mr Lewis had $360. He spent ⁷/₉ of it.
How much money did he have left?

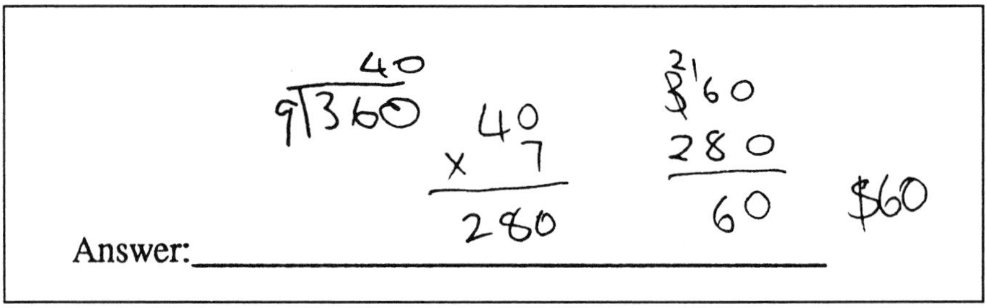

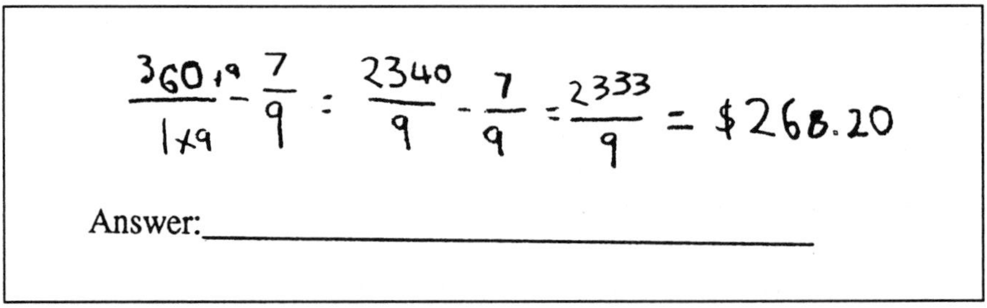

Figure 13.7. Example of a Population 2 Problem-solving Item with which Australian students had difficulty

6.   IMPLICATIONS FOR MATHEMATICS
TEACHING

The evidence of areas of weakness sketched in this chapter led to a call from senior university mathematicians for action to be taken to improve the mathematical skills of Australian students at primary and secondary levels. For example, in an interview for *The Adelaidian* (the newspaper for the University of Adelaide), the Dean of the Faculty of Mathematical and Computer Sciences, Professor Alan Carey, said that mathematicians had been concerned for many years with the impact of the way that mathematics is taught at school level. In arguing for a review of current practice, Carey explained:

> Our contention has always been that mathematics cannot be taught in the same way as, say, history or geography. Mathematics ... shares with music the need to practise certain skills (and) a sequential structure. For example, it is not possible for someone to sit down at a piano and play a complicated piece without long and sometimes arduous practice. ... For many years educationists have argued that in order to make mathematics accessible we need to minimise the skill development and teach the ideas. The musical equivalent is learning about the piano rather than how to play the piano. (Ellis, 1997, p. 3).

Professor Carey went on to suggest that there are three mistaken views that, if eliminated, 'would go a long way to fixing the problem' of Australian students' weaknesses in mathematics: that skill development is not important; that things which average students find hard should not be taught; and that specialist training in mathematics is not necessary for those who teach it at primary and early secondary levels.

Speaking at a national conference on the implications of TIMSS for mathematics and science education in Australia, Professor Stacey from the Science and Mathematics Education Department at the University of Melbourne said:

> My assessment is that Australia has gone down in one area (computation) and this does not of itself concern me. However, Australia has not seen a corresponding improvement in basic understanding, which might have been expected even in a topic as central to everyday life as number. (Stacey, 1999, p. 45)

In her paper, Stacey proposed several strategies to help students achieve better understanding of mathematics, which she believes could be done without an allocation of extra time:

- to start earlier on some key ideas and aim to do them more thoroughly;
- to increase the conceptual demand of activities commonly used for developing understanding;
- to promote teaching practices that really focus on students' learning;
- to expect high achievement and include challenging work in the curriculum;
- to engage in serious construction of sequences of lessons to develop stronger understanding in the longer term; and
- to take a careful, professional, long-term approach to state and national curriculum development.

7.   CURRICULUM ASPECTS

Both Carey and Stacey were united in advocating a review of current mathematics teaching at school level, though from somewhat different perspectives.   In view of their comments that knowledge and routine procedures have been de-emphasised in recent years, the finding from the TIMSS curriculum and textbook analyses that textbook space is largely concentrated on these matters is paradoxical.   Results of these analyses showed that Australia was evenly balanced in the numbers of mathematics topics introduced at least three years earlier or later than the median grade level, as distinct from the half dozen or so countries which introduced more topics later and the majority of countries which introduced more topics earlier.   On average, Australia retained topics in the curriculum for about a year longer than the median number of years (Schmidt *et al.*, 1997).

Some further analyses in Australia, which were commissioned following the students' performance in TIMSS, have provided a much more detailed picture of when mathematics topics are introduced into the curriculum at primary and early secondary levels.   These analyses help to explain why Australian students could not match the achievement of students in the Asian countries which are our major trading partners.   Statements of curriculum expectations from those countries, plus from other countries such as the Czech Republic, Hungary and some of the Canadian provinces, were carefully reviewed.   With few exceptions, these reviews revealed a picture of topics expected to be mastered at an earlier stage in other countries than has been proposed for Australia in the recently developed numeracy 'benchmarks' (Lokan & Ainley, 1998).

While the intention of the benchmarks is to define a minimum acceptable standard of achievement without which students would have difficulty progressing in their schoolwork (McLean, 1997), the results of the comparisons undertaken, together with the TIMSS curriculum analyses, both point to the desirability of a revised curriculum. The mathematics curricula now in place in most states were not introduced until the mid-1990s and are somewhat different from the curricula analysed in 1992-93 for TIMSS. Already, however, two states are currently reviewing theirs, and the TIMSS results were one of the factors in the decisions to undertake the reviews.

REFERENCES

Ellis, D. (1997) Maths study shows cause for concern. *The Adelaidean* November, vol 6, no 22, pp. 1-3 (University of Adelaide)

Keeves, J. P. (1992) *Learning Science in a Changing World. Cross-national Studies of Science Achievement: 1970 to 1984* (The Hague: International Association for the Evaluation of Educational Achievement)

Lokan, J., Ford, P. & Greenwood, L. (1996) *Maths and Science on the Line: Australian Junior Secondary Students' Performance in the Third International Mathematics and Science Study.* TIMSS Australia Monograph No. 1 (Melbourne: Australian Council for Educational Research)

Lokan, J. & Ainley, J. (1998) The Third International Mathematics and Science Study: Implications for the development of numeracy benchmarks. *Unicorn*, 24, pp. 97-109

McGaw, B., Long, M., Morgan, G. & Rosier, M. (1989) *Literacy and Numeracy in Victorian Schools: 1988.* ACER Research Monograph No 34 (Melbourne: Australian Council for Educational Research)

McLean, K. (1997) Towards a National Report Card. *Education Quarterly*, No. 3, Spring, pp. 5-7

Rosier, M. J. (1980) *Changes in Secondary School Mathematics in Australia: 1964–1978.* ACER Research Monograph No. 8 (Melbourne: Australian Council for Educational Research)

Schmidt, W. H., McKnight, C. C., Valverde, G. A., Houang, R. T. & Wiley, D. E. (1997) *Many Visions, Many Aims: A Cross-National Investigation of School Curricula in Mathematics* (Dordrecht: Kluwer)

Spearritt, D. (1987) Educational Achievement. In Keeves, J. P. (ed.), *Australian Education: Review of Recent Research* pp. 117-146 (Sydney: Allen & Unwin)

Stacey, K. (1999) Implications of TIMSS for Mathematics Education, in Lokan, J. (ed.), *Raising Australian Standards in Mathematics and Science: Insights from TIMSS.* Proceedings of ACER National Conference 1997 pp. 117-146 (Melbourne: Australian Council for Educational Research)

Chapter 13

TIMSS in South Africa
The Value of International Comparative Studies for a Developing Country

S J Howie
Education and Training Systemic Studies, Human Sciences Research Council, Pretoria, SA

1.   BACKGROUND

It has been said that 'we all learn from comparisons, and the history of comparing and borrowing ideas in education' (Brickman, 1988). There has been growing interest in developing a systematic approach to assessing outcomes in education. Concerns are growing in governmental and education circles everywhere that many children spend a considerable amount of time in school, but acquire few skills. Governments are faced with expanding enrolments, while at the same time improving the quality of education without increasing expenditure. Both of these issues affect developed and developing countries.

These issues have led to an increasing number of studies and calls for monitoring systems to be introduced in many parts of the world. These studies have focused on describing and monitoring the nature of students' achievement, the relevance of these achievements to the world of work, and the number of inadequately prepared students leaving the education system (Greaney and Kellaghan, 1996).

Comparative studies of pupils' achievement have been used to gauge the relative status of countries in developing individual skills. The development of computer and information technology since the 1960s has also served to encourage large-scale international assessments since they made large-scale measurement, control and analysis of influences on school achievement possible. For instance, the development of IEA methodologies meant that its studies not only collected achievement scores, but also constructed and

D. Shorrocks-Taylor and E.W. Jenkins (eds.), Learning from Others, 279–301.

applied attitude scales and background information questionnaires which could be used to explain differences in achievement outcomes (Postlethwaite, 1988). International comparative studies have also resulted in a number of debates, with researchers and policy makers often debating the methodological issues, rather than the more painful educational realities highlighted by the studies.

1.1   Developing countries participating in international studies

The importance of international studies for developing countries cannot be overlooked, despite the difficulties and problems of conducting them. They provide a comparative framework in assessing student achievement and curricular provision. They also give an indication of where students in a country are relative to other students in another country, and they show the extent to which the treatment of common curriculum areas differs across countries.

Findings of international studies have been used to change educational policy, of which examples are Canada, Japan, Hungary and the USA (Greaney and Kellaghan, 1996). They attract more media coverage than national studies and this results in national debates and often changes on a local or provincial level. The national team managing the studies gains valuable experience, which may contribute greatly to the development of local capacity to conduct research and national assessments.

International studies have also highlighted the differences in achievement, teaching and learning in different subjects between developed and developing countries. For instance, Heyneman (1986) reported that some of the international studies suggested that in many less-developed countries little relationship exists between socio-economic status and educational achievement, which is contrary to the findings in developed countries. Lockhead (1988) suggested that educated parents in developing countries are more likely to provide learning in the home and other resources and to send their children to schools with better resources and better trained teachers.

Many researchers in developing countries are often isolated from the academic and intellectual processes taking place in developed countries, although the advent of the Internet is assisting to reduce this isolation. International studies offer them the opportunity to study other systems within a well-designed research model and to build the capacity to undertake the research independently at a later stage. Cross-national research also widens our horizons and can suggest non-incremental means of improvement (Finn, 1989).

From a developing country's perspective, these studies provide benchmarks for educational performance. They inform curriculum developers of what is happening in other countries and provide examples of success and failure. An important function of international comparative projects is that they generally indicate areas within subjects or schooling that need attention. The information generated helps to fuel efforts to strengthen education. Although there are areas that are more difficult, or that some would say are impossible to measure 'we must not forget the immeasurable, but we must not slight what can be measured' (Finn, 1989).

This is an important sentiment which critics of international studies often lose sight of. For a developing country like South Africa, these studies provide the kind of information that can inform the education departments whether the current efforts are productive and which areas need attention.

1.2 Tensions around international studies

Over the past two decades, a number of tensions concerning international studies have developed. Questions have been raised concerning the use of inputs, in particular that they are difficult to determine and that the limited information collected results in explanations that are questionable.

There are concerns that comparative studies are often used and abused by policymakers, and political rhetoric often follows the release of international studies results. The concern is that policy makers will select those indicators that they see as relevant within a political agenda (and not an educational one) when they propose educational reform. However, this argument and concern could also apply to results from national studies.

Rotberg (1991) argues that factors such as the availability of schools and materials, opportunity to learn, status and quality of teaching, parental interest, and class size differ so radically from country to country that valid comparisons of international achievement results are impossible. However, Heyneman (1986) counters this stating that 'many educational phenomena are similar from country to country and from culture to culture and that concerns for differences should not obscure the similarities'. Keeves (1992) also points to statistical and measurement problems in these studies. However, he also maintains that the data should be judged on the strength of and the meaningfulness of the relationships that these studies report.

A potentially significant problem that is often at the heart of the tension in these studies is the difficulty of sampling. Developing countries especially have a disadvantage due to up-to-date population data not being available and communication and logistical problems resulting in relatively low response rates. There is also a general concern about the difficulty of developing a test that is equally valid for several countries due to the different curricula used.

Developing countries have additional, but common, problems concerning their participation in international studies. These include:

- the unavailability of information on schools and enrolment figures;
- the lack of experience in administering large-scale assessments/objective tests to schools;
- tests do not adequately reflect the curriculum offered in schools or do not reflect regional, ethnic and linguistic variations;
- lack of exposure to objective-type items;
- fear that test results might be used for teacher accountability purposes;
- insufficient funds and skilled manpower to do rigorous in-country analyses of the national and international data;
- governmental restrictions on publicising results; and, finally,
- logistical problems in conducting the assessment.

2.   THE SOUTH AFRICAN CONTEXT

One of the biggest challenges facing South Africa at present is the provision of quality education for all its people. Education is the key to a genuine participatory democracy and the foundation for the achievement of sustainable development in any country.

South Africa's 41.5 million people form a multi-cultural society and the population consists of four ethnic groups: Indian (1 million people), coloured (3.5 million people), white (5.4 million people) and African (31.6 million people) (Central Statistical Services (CSS) 1995). There are eleven dominant languages spoken that have all been declared official languages of the country. In practice, English (which is spoken by 9.1 per cent of the population) is the language of business and government (replacing Afrikaans). However, it is not the most widely spoken language at home.

South Africa is an economically developing country where first and third worlds meet. The majority of the population still resides in typical Third World impoverishment, while state-of-the-art technology in manufacturing and mining (in particular) is being developed and utilised by South Africa's scientists and engineers. Some of the significant challenges facing South Africa are discussed below.

2.1   World competitiveness in a global market

In 1995, South Africa was placed 42 out of 48 countries in the annual World Competitiveness report of the Economic Forum and the Harvard

Institute for International Development. In particular, South Africa was placed 48/48 for the development of human resources for a competitive economy. In 1997, South Africa was placed 44 out of 53 countries, below countries like Brazil and Colombia and was again placed last with regard to the development of human resources. In the 1998 Africa Competitiveness report, South Africa was placed 7th, behind Mauritius, Tunisia, Botswana, Namibia, Morocco and Egypt.

2.2   Unemployment and economic growth in South Africa

More than nine million South Africa children live in poverty-stricken homes (Howie, 1998). The South African Institute of Race Relations' survey of 1995/6 calculated an unemployment rate of 32.6 per cent of the economically active population of 14.3 million South Africans, with the rate being 41.1 per cent for the African population. It is estimated that in 2010, there may be eight million unemployed people and a shortage of 200,000 skilled workers. It is further estimated that seven out of every 100 people who seek employment will find jobs (Gouws, 1997). In order to attempt to rectify this situation, South Africa needs a real GDP of 6.1 per cent by the year 2000 in contrast to the current figures of 2.0 per cent and 2.5 per cent (van Eldik, 1998).

2.3   Education in South Africa

South Africa's political history is well-known. Its impact on the education system and the youth that have passed through this system is especially devastating. The illiteracy rate of 55 per cent amongst South Africa's disadvantaged communities proves this (Gouws, 1997). South Africa is now divided into nine provinces incorporating former provinces and black homelands.

The present education system is an amalgamation of 17 different education departments that were merged in 1995 as a result of the change in government policies. Under-resourced and often mismanaged, many of these departments provided insufficient and ineffective education to millions of young people, especially from the African community. The education system is still grappling with the inequalities of the apartheid era whilst trying to maintain the standard of education received by the white pupils in previous years.

South Africa has two different types of schools: government and private schools. In total there are more than 32,000 schools of which the majority are comprehensive (offering general education) in nature. Although schooling is compulsory up to the end of Grade 7, absenteeism is rife and there is also a high drop-out rate throughout the schooling.

Conditions in many South African schools are such that many in first world countries would scarcely believe them. In 24 per cent of the country's schools, there is no running water within walking distance. Sixty-seven per cent of the schools have no electricity and in most provinces between 50 per cent and 80 per cent of the schools have no telecommunications. Thirteen per cent of the schools have no toilets. Many schools have a serious shortage of classrooms and, of these, many are in an uninhabitable condition. The student to teacher ratio is also very high, and in three provinces was found to be more than 40:1. It is not uncommon to find classes of more than 100 pupils being taught by one teacher in classrooms designed for 30 students (HSRC, 1997).

The government has been working with these difficulties since the 1994 election. A number of government papers have emerged, including the White Paper on Education, highlighting the government's awareness of its inheritance as well as intended policies to bring about change. These papers marked the beginning of a new era in education in South Africa. For instance, education and training are combined in all of the education department's documents, in recognition of the importance of training in schools, teacher training colleges and higher education institutions. It is also seen as a move towards integrating the academic and practical, theory and practice, knowledge and skills, and the mind and the hand.

The ANC policy framework document (1994) described the situation in science education as the following: 'Science and mathematics education in African schools in South Africa is characterised by a cycle of mediocrity'. 'The infrastructure for the teaching of science and mathematics is poor, especially at senior secondary level. Materials are in short supply. Most schools lack laboratories. Teachers are under-qualified. In African colleges of education science and mathematics are low status subjects only taught at matriculation level to diploma students. Under-qualified and poorly prepared teachers in turn produce weak and poorly prepared school students'. 'The cycle of mediocrity is reinforced by the unsuitable nature of the science and mathematics curriculum in the schools. The curriculum is academic, outmoded and overloaded. Applied science and technology as well as the social and ethical aspects of science are excluded. The consequences of a lack of a suitable curriculum cut across racial divide. Only 12 per cent of higher education students pursue degree and diploma programmes in engineering and the life, physical and mathematical sciences'.

3.   IMPLEMENTING TIMSS IN SOUTH AFRICA

3.1   Previous indicators of pupils' achievement

The matriculation examination (the external final examinations written at the end of Grade 12) figures have been used as an indicator of South African students' performance at school level. The results at the end of 1997 were the lowest recorded since 1979. Whilst the number of candidates writing these examinations increased by 7 per cent from 1996, the percentage of failures increased by 23 per cent. The inefficiency of the govern-ment school sector, in particular, results in resources being applied to pupils who do not pass through the system successfully. This is problematic as most of these schools have very scarce resources. In an historical overview of school enrolments and success rates at school, de Villiers (1997) quotes very disturbing figures. When analysing students from different ethnic groups and their progress through the schooling system, he found the following:

– 69 out of every 100 Grade 1 White students complete their matriculation examination within 12 years;
– 62 out of every 100 Grade 1 Indian students complete their matriculation examination within 12 years;
– 19 out of every 100 Grade 1 Coloured students complete their matriculation examination within 12 years;
– 8 out of every 100 Grade 1 African students complete their matriculation examination within 12 years.

One of the issues clearly emerging is that these high failure rates indicate that there is a problem in the junior phases of the education system. Certainly pushing students up into a higher grade without them seemingly being able to cope with that year's work is resulting in very high failure rates in the final year of schooling.

The South African White Paper on Education singles out mathematics and science as important school subjects and recognises the importance of more students leaving school proficient in these subjects. However, from Grade 10 onwards physical science and mathematics become optional subjects.

The situation regarding the enrolments in mathematics and science and the matriculation exemption with these subjects also varies dramatically between ethnic groups. For instance, whilst only 1 in 312 African students entering the school system leaves with physical science and mathematics as final year subjects, 1 in 5.2 White students, 1 in 6.2 Indian students and 1 in 45.9 Coloured students obtains a matriculation exemption with physical science and mathematics (Blankley, 1994).

The primary reason for the Human Sciences Research Council (HSRC) undertaking TIMSS in South Africa was to use it as a longitudinal study. It was intended to monitor the changes taking place in mathematics and science education in South Africa during and after the transition of the new government, and the reforms that have been introduced into the South African education system. These include the new curriculum, Curriculum 2005, an outcomes-based model that is currently being implemented in its pilot phase. The HSRC felt that TIMSS would be useful in monitoring trends in teaching and pupils' achievement in mathematics and science on a national and provincial scale. It was also felt that the curriculum-based research undertaken as part of the TIMSS project would be useful in order to compare the South African curricula with those of other countries.

3.2   Students participating in TIMSS in South Africa

TIMSS focused on 9 year-old students, 13 year-old students (Grade 7 and Grade 8 in South Africa) and final year secondary school (Grade 12 in South Africa) students. However, South Africa participated in only the latter two groups. Students in South African schools learn in their mother tongue/ home language until Grade 5 when the language of instruction becomes English or Afrikaans. Translating the TIMSS test booklets into another nine languages was not viable at the time of testing and testing these very young children in English or Afrikaans did not seem to be an option given that they had only received a couple of months instruction. The decision was made to exclude the first population group of 9 year-old students. A nationally representative sample yielded 300 schools for the 13 year-old age group. This included 150 primary schools and 150 secondary schools as two grades containing the most 13 year-old students had to be tested. The final-year students' (Grade 12) sample was the same as the secondary schools that were used for the 13 year-olds (150 secondary schools). South Africa participated in the general science and mathematics literacy test, rather than the specialised mathematics and physics tests. The reason for this was that the mathematics and physical science curriculum followed by South African schools at the final year level is not as specialised as that of other countries. Therefore, it was felt that it would be inappropriate for South African students to write these specialised tests. Furthermore, it was also relevant for researchers to try to ascertain the mathematical and scientific literacy of the general population of students in their final year.

A national sample of schools, fully representative of provinces, ethnic groups, and urban and rural communities, was randomly selected. This sample consisted of 14,020 students, with 5,532 students in Grade 7, 4,793 students in Grade 8, and 3,695 students in Grade 12. The results presented in the national and international reports (after the data cleaning process) reflect the achievement of 12,437 South African students, of whom 5,267 were in Grade 7 at 137 schools, 4,413 in Grade 8 at 114 schools and 2,757 Grade 12 students at 90 schools.

3.3   TIMSS Project challenges

The challenges facing the TIMSS project team were many and varied. Although all countries participating in large-scale surveys are faced with difficulties, conducting these studies in a developing country posed unique challenges. It was the first time that a survey of this size had been carried out in science and mathematics education in South Africa. The approval for the project had been delayed and this led to the situation where the researchers were under constant pressure for the remaining years of the project to catch up. The result was that not all the detailed procedures could be undertaken. Instruments were piloted in eight South African schools, but there was no time to analyse the data collected. Not following all these procedures resulted in several difficulties when the study was implemented on a full scale. Training courses were conducted for the researchers working on the team to build the necessary capacity to undertake the project. The fieldwork in itself was the biggest challenge. A division of the HSRC specialising in the collection of data was contracted to collect the data.

The challenges in conducting the study in South Africa arose from:
- lack of information about schools, due to incomplete lists of schools or departmental staff not co-operating;
- difficulty in communicating with the schools, due to no telecommunications or inefficiency within schools;
- the isolated situation of schools (and even schools that could only be reached on foot);
- the fact that at this time the provincial boundaries were being redefined;
- the presence of schools in the sample that did not exist, a phenomenon in South Africa called 'ghost' schools;
- inefficient postal service that resulted in a large number of lost letters and returned questionnaires;
- the geographic size of the country;
- vehicle breakdowns, poor roads and/or lack of roads and accidents;
- the lack of co-operation of some schools and/or staff members being absent without reason.

As can be seen, many of these challenges apply generally to developing countries. Most of these were addressed by persistence of the field-workers and the national centre project staff, for instance some schools were visited up to four times in an effort to secure testing times and to test the students. Replacement schools were also used in cases where even this persistence was inadequate. The assistance of the provincial authorities was invaluable in locating schools and enabling the field-workers access to schools. The experience of many of the field-workers (in these type of conditions) resulted in the relatively successful response rate of 91 per cent for Grade 7 classes sampled, 76 per cent for Grade 8 classes sampled and 60 per cent for Grade 12 classes sampled (including replacement schools).

4.   TIMSS RESULTS IN SOUTH AFRICA

The broader community received the South African TIMSS students' results with shock and dismay.

The overall results for Grades 7 and 8 and for Grade 12 students, shown in Tables 12.1, 12.2, and 12.3 were very low in comparison with other countries (and in fact were the lowest of all countries participating). This was also true for the results in mathematics and science separately for all three grades.

Table 12.1. South African Grade 7 and Grade 8 students' results for science compared to other newly developed or developing countries

COUNTRY	GRADE 8	GRADE 7
Singapore	607 (5.5)	545 (6.6)
Iran	470 (2.4)	436 (2.6)
Colombia	411 (4.1)	387 (3.2)
Thailand	525 (3.7)	493 (3.0)
South Africa	326 (6.6)	317 (5.3)
International mean	516	479

Source: Beaton et al (1996b)
( ) Standard errors appear in brackets in the table

As can be seen from Table 12.1, South African students' performance in science fell well below that of students from other countries. The South African science curricula were not comparable with those of the other countries. Only 18 per cent of the items were related to content included in the Grade 7 science curriculum and 50 per cent of the items to the Grade 8 science curriculum.

Table 12.2. South African Grade 7 and Grade 8 students' results for mathematics compared to other newly developed or developing countries

COUNTRY	GRADE 8	GRADE 7
Singapore	643 (4.9)	601 (6.3)
Iran	428 (2.2)	401 (2.0)
Colombia	385 (3.4)	369 (2.7)
Thailand	522 (5.7)	495 (4.8)
South Africa	354 (4.4)	348 (3.8)
International mean	513	484

Source: Beaton et al (1996a)
( ) Standard errors appear in brackets in the table

The South African students' performance was a little better in mathematics, but still below the international averages of the other countries. Furthermore, substantially more of the mathematics items were found in the mathematics curricula. In total, 50 per cent of the items were found in the Grade 7 mathematics curriculum and 80 per cent of the items in the South African Grade 8 mathematics curriculum.

Table12.3. South African Grade 12 students' results for mathematics and science literacy compared to selected countries

COUNTRY	OVERALL LITERACY	MATHEMATICS LITERACY	SCIENCE LITERACY
Netherlands	559 (4.9)	560 (4.7)	558 (5.3)
Sweden	555 (4.3)	552 (4.3)	559 (4.4)
Canada **	526 (2.6)	519 (2.8)	532 (2.6)
New Zealand	525 (4.7)	522 (4.5)	529 (5.2)
Australia **	525 (9.5)	522 (9.3)	527 (9.8)
Russian Federation	476 (5.8)	471 (6.2)	481 (5.7)
Czech Republic	476 (10.5)	466 (12.3)	487 (8.8)
USA **	471 (3.1)	461 (3.2)	480 (3.3)
South Africa *	352 (9.3)	356 (8.3)	349 (10.5)
International Mean	500	500	500

Source: Mullis *et al.* (1998: 32-48)
( ) Standard errors appear in brackets in the table.
* Unapproved sampling procedures and low participation rates
** Did not satisfy guidelines for sample participation rates

South Africa was the only industrially developing country in the Grade 12 study from Africa, Latin America or Asia. The performance of South Africa's top performing students did not compare with the top performing students of other countries. The fact that the top performing students from SA performed, on the whole, at the same level as the average students from other countries, was both surprising and disappointing to educationists in South Africa.

4.1 Findings from achievement and literacy tests

The difference in the results for South African students in Grade 7 and Grade 8 was minimal. The difference represented the lowest increase in result from the lower grade to the upper grade of all the countries taking part. Furthermore, South African students in Grades 7 and 8 and Grade 12 did not excel in any individual mathematics or science topic in the TIMSS achievement and literacy tests and the results were uniformly poor. The performance of the final year (Grade 12) students varied considerably across different provinces in South Africa. Students from the two most economically impoverished provinces, which also have the greatest number of under-resourced schools, performed worst in the final year students' science literacy test and overall. An unexpected finding was that South Africa was the only country with no significant difference between the performance of boys and girls in Grade 12 and one of a couple of countries exhibiting no gender difference in Grades 7 and 8.

The performance of Grades 7 and 8 students in the mathematics and science questions based on the curricula was the same as in those that were not in the curricula, which was somewhat surprising. While, as noted above the mathematics curricula were comparable with the international curricula, this was certainly not the case for the science curricula (especially for Grade 7).

The Grade 12 South African students who did not take science at school did not score significantly less in the science literacy test than other groups of students taking mathematics and science at school. In fact the highest scores from this 'non-science' group of students exceeded the national average and were comparable to the international average.

South African students in both studies struggled with the free-response questions and appeared unable to formulate their own answers. South Africa was the only country where the language of the test was not the home language of the majority of students. There was evidence amongst the students from all three grades of language difficulties and approximately 80% of all the students wrote the tests in their second language. Grade 12 students with English or Afrikaans as their home-language (the language of instruction and hence the language of the test) performed better than those speaking another language at home.

4.2   Findings from students' questionnaires

In both studies, students were older than their international counterparts in TIMSS. In Grade 7, South African students were 13.9 years and in Grade 8 they were 15.4 years compared to the international averages of 13.2 years and 14.2 years. The South African Grade 12 students were 20.1 years of age compared to the international average age of 18.6. Given the high numbers of students repeating grades in South African schools, this was no real surprise. In the Grade 12 sample, more than 16 per cent of the students were repeating Grade 12 and yet they were still in a mainstream school.

From the international TIMSS results, it was clear that parents' education was positively linked to students' mathematics and science literacy. Parents' higher education levels were related to students' higher literacy. South African students' parents' education levels were found to be very low. More than 30 per cent of the South Africa students' parents (as with Australia, Cyprus, Czech Republic, France and Italy) had completed primary school only. In South Africa, a symptom of this was that the average number of books in the homes of students in both studies was substantially less than the international average.

At schools, South African Grade 7 and 8 students had less learning time than students in the top-performing countries and the amount of homework given to Grade 7 and 8 students was much lower than the international average. Students in Grades 7, 8 and 12 indicated biology as their preferred science subject, with chemistry at the bottom of the list of choices.

5.   REACTION TO TIMSS RESULTS IN SOUTH AFRICA

The media attention given to the results of TIMSS generated a lot of interest in education and political circles as well as amongst leaders in the scientific and technological fields. What received most attention was that top-performing South African students simply were not comparable to the top-performing students of other countries.

As a result of the attention, several initiatives have been taken. The TIMSS results have been discussed at cabinet level where opposition party politicians have placed pressure on the Education minister to follow up the findings of TIMSS and to launch new initiatives. In particular, the Education Minster stated during a Parliamentary debate that 'aspects that will receive particular attention are: 'addressing the mismatch between South Africa and international curricula', 'addressing the structure of text-books', 'investigating the use of innovative methods and classroom approaches' and further aspects to be attended to are, 'acquiring information on the syllabuses and methodologies on the seven best countries, comparing their educational systems and curricula, and extracting the best to be implemented in South Africa', 'formulating a national mathematics, science and technology vision', and 'promoting mathematics, science and technology education among girls'. (National Assembly, 1997, pp. 111-114).

TIMSS also added weight to the implementation of the Year of Science and Technology (YEAST) during 1998. The Department of Arts, Culture, Science and Technology arranged YEAST together with the Science Councils, other government departments and the private sector. National initiatives at provincial and local level were organised to generate public awareness and interest in scientific, engineering and technologically related fields.

Finally, the Department of Education, disturbed by the mature age of the students highlighted by the TIMSS' findings, launched an enquiry into this disturbing trend.

6.   KEY ISSUES HIGHLIGHTED BY THE TIMSS PROJECT IN SOUTH AFRICA

South Africa was isolated from the rest of the political and academic arena for a long time and therefore to a large extent had fallen out of the international education and other debates. The participation in TIMSS rejuvenated the academic and policy debates in mathematics and science education. There has been much talk in government circles about the importance of mathematics, science and technology and about increasing numbers of students going into these subjects. Nevertheless, there still appears to be inadequate funding for school science, technology and mathematics subjects as well as the inadequate provision of facilities and equipment at many schools. TIMSS focused the country's attention on the poor overall performance of South Africa's students, but also indicated several key issues that will be discussed in this section.

6.1   South Africa's Outdated curricula

The TIMSS project through the test-curriculum match analysis and the curriculum analysis study revealed that the South African Grade 7 and Grade 8 science and mathematics curricula were very different from those of the other countries in the study. In particular, there was a marked difference in the Grade 7 curricula for both subjects.

These results substantiated the concern about South Africa's curricula that had originally led to the present government to introduce Curriculum 2005. Curriculum 2005 aims to 'develop citizens who are active and creative, inventors and problem-solvers, rather than dependent and unthinking followers and also to inculcate an appreciation for diversity in the areas of race, culture and gender'. Curriculum 2005 is seen to be the boldest attempt at curriculum renewal in South Africa's recent history and aims to align the schools with the demands of the workplace and the social and political aspirations of South Africa's citizens. Firstly, the contents of the curriculum are being re-written in terms of clearly defined outcomes and secondly there is a planned shift in the focus of classroom activity to student activity and collaborative learning.

6.2   Ineffective schooling system

One of the findings from TIMSS was the mature age of the students in each of the Grades tested. In two of the three grades, the South African students were the second oldest of all the participating countries, and the oldest in the third grade. This finding is supported by de Villiers' work on the inefficiency of the South African schooling system and the extent of the repetition of grades taking place in the South African schooling system.

6.3 Learning culture and attitudes to mathematics and science education

The attitudes of the South African students towards mathematics and science were generally positive for all three grades, perhaps more so than expected. It was also interesting to find that students in all three grades felt that they normally performed well in these subjects contrary to their results in TIMSS. TIMSS also revealed that biological science was the preferred subject for all three grades. For instance, 32 per cent of the Grade 8 students chose biology as their preferred science subject compared to 24 per cent for physics and 7 per cent for chemistry (Howie, 1997). Similar trends were found for Grades 7 and 12.

6.4 Commitment of resources to science and mathematics education

TIMSS also highlighted the lack of resources for the South African schools. These include the lack of physical science laboratories and mathematics and physical science textbooks. As mentioned previously, many schools are without the basic facilities such as toilets, running water, electricity and telecommunications. Furthermore many schools are without textbooks, classrooms, libraries and laboratories, making science teaching, in particular, very difficult. With up to 90 per cent of the education budget being spent on the costs of teachers' salaries (Bot, 1998), large-scale retrenchments are currently underway in order to find money for the provision of equipment, textbooks and facilities. The fact that expenditure on textbooks (in general) decreased by 41 per cent between 1995 and 1996 (Bot, 1998) is alarming, especially for physical science. In areas where there are no facilities like libraries, both students and teachers rely heavily on the textbooks as their sole resource in this subject (and others). As a result of the cuts in the expenditure on textbooks, many students will go through school with no access to learning resources or materials.

6.5 Language, communication and science and mathematics

TIMSS revealed the significant language and communication problems of South African students learning mathematics and physical science in a second language. Students in all three Grades showed a lack of understanding of both mathematics and physical science questions and an inability to communicate their answers in instances where they did understand the questions. Students performed particularly badly in questions requiring a written answer. Clearly much attention is required to address this language and communication problem. Special programmes will need to be investigated and introduced to assist students and teachers in overcoming this.

6.6   Disparities within the education system

The TIMSS results highlighted the disparities in the schooling systems with students in schools from traditionally disadvantaged areas achieving lower scores than students from schools in more advantaged areas, although exceptions were found for individual scores. The differences in achievement cannot be easily explained given the complexities embedded within these disparities. They can be partly explained by the unequal provision of resources and facilities in the past; different socio-economic backgrounds; and the variation of students' home languages, 80 per cent of which differed from the language of the test.

6.7   Ineffective tuition

No significant difference was found between Grade 7 and Grade 8 students' achievement in mathematics and science. This finding raised concerns about the effectiveness of schooling and the curriculum in these grades.

One other finding that questions the effectiveness of schooling was that there was little difference in the level of mathematics and science literacy between students who are enrolled in a physical science course and those students who do not take any science or mathematics course.

Several of the issues are related to teacher competency. There is a major concern amongst educators in South Africa that the under-qualified or unqualified teachers will not be able to implement the new curriculum in the year 2005. South African teachers are suffering from low morale related to many factors including the uncertainties associated with the changes and reform in education, low and delayed salaries, threat of retrenchment, re-deployment, lack of confidence, unmotivated and ill-disciplined students, no support staff and a lack of teaching materials.

Currently there is an insufficient supply of adequately trained mathematics and science teachers. In particular, if one looks at science, less than 50 per cent of science teachers have accredited training in science. Aggravating the situation, there is also a very high attrition rate as about 2,100 science teachers leave the teaching profession each year.

6.8 Unrealistic expectations of students

A very high percentage of students (75 per cent) indicated that they intended going to university after school, which appears to be completely unrealistic, given the generally low levels of mathematics and science literacy and achievement at school. However, what it also indicates is the lack of career guidance from schools as well as the over-emphasis on university-type education at the expense of more suitable occupations for students of differing abilities. It may also reflect the ambition of the majority of people who have been politically and economically oppressed for decades.

7. THE WAY AHEAD

The key issues discussed in the previous require urgent attention. In this section several points are discussed directly related to addressing the challenges facing mathematics and science education in South Africa.

7.1 Future international studies in South Africa

The Human Sciences Research Council (HSRC) is co-ordinating the Third International Mathematics and Science Study-Repeat (TIMSS-R). This is essentially a repeat of TIMSS, but focusing only on the Grade 8 students' achievement in mathematics and science. In South Africa, a sample of 225 schools has been drawn across the nine provinces (25 schools per province) and the data was collected in October and November 1998. The data will be analysed for trends identified in TIMSS in 1995. Planning for this study includes the challenge of avoiding some of the obstacles that occurred in TIMSS previously, whilst trying to anticipate new ones.

The HSRC is also co-ordinating the Second Information Technology in Education Study (SITES) in South Africa. Only ten per cent of schools in South Africa are known to have some form of Information and Communications Technology and therefore a smaller sample of schools will be drawn. The study will be conducted at approximately the same time as TIMSS-R.

7.2 Monitoring changes in science and mathematics education

A critical analysis as well as a monitoring system will have to be put in place to keep track of the development and implementation of Curriculum

2005, and other envisaged changes in science and mathematics education, to ensure the most successful implementation possible. This is a role that the HSRC could play with projects such as TIMSS, TIMSS-R and SITES, in close collaboration with the Department of Education.

7.3   Government policies and interventions in science and mathematics education

The White Paper on Science and Technology (1996) published by the Department of Arts, Culture, Science and Technology emphasises the importance of an awareness and understanding of science, engineering and technology (SET). The importance of a national campaign to promote the public understanding of SET is seen to be an integral part of the government's plan to close the gap between the first world and the third world segments of South Africa's population.

The importance of science, engineering and technology awareness so explicitly stated in these papers emphasises the need for the successful implementation of a National System of Innovation (NSI) in South Africa. The NSI is a system whereby knowledge, technologies, products and processes must be converted into increased wealth and an improved quality of life for all South Africa's citizens. The process by which this transformation of new knowledge into new wealth takes place is generally known as (technological) innovation. The aims within the White Paper are such that only with the close collaboration of the Department of Education can they be realised successfully.

South Africa, since 1994, has developed policies in both Education and Science and Technology. However, to date there has been no close co-operation between the two departments (Department of Education and the Department of Arts, Culture, Science and Technology) producing these policies. This is essential for the technological and industrial development of South Africa and the region. The linking of these policies will surely influence the curriculum reform in science and mathematics education in South Africa.

7.4   Curriculum reform in science and mathematics education in South Africa

Innovation in South African education at present is heavily influenced by the experience of the more-developed countries, especially the United Kingdom, United States of America, Australia and Canada and there are frequent visits to and from these countries by educationists.

What is interesting to observe, using Bloom's four published suggestions for improving school instruction and curriculum, is that the South African government has chosen the most expensive option. Bloom stated that the 'most costly approach (to reform education) is to embark on a major reform of the curriculum. The success of this depends very much on the availability of highly trained curriculum specialists, who carefully weigh up the effectiveness of each new step as it is implemented. Success depends on the thorough retraining of existing teachers who are going to be involved in teaching according to the new curriculum'. Curriculum 2005 appears to be driven largely by a political agenda and therefore the time pressures are great on the implementers within the Department of Education.

7.5   The preparation of science
and mathematics teachers

One of the key issues highlighted by TIMSS was the ineffectiveness of the schools in South Africa. One of the problems directly linked to this is the question of teacher preparedness. The pre-service training of teachers is in a state of flux as the colleges of education are being reshaped and the curricula in teacher education programmes (at colleges of education, technikons and universities) nationally are being reviewed in preparation for the new school curricula. Whilst many colleges are being transformed into Agricultural, Nursing and Community colleges, several have been merged with other colleges or closed. However, there is a question mark over the quality of teachers being produced from these colleges and careful attention will need to be paid to these to improve the quality of science and mathematics teachers emerging. In particular, the science content knowledge of students coming into the colleges is very weak and often does not meet the expected minimum level of knowledge required in their first year (Howie, 1998b). In an attempt by the government to address the question of quality of science and mathematics educators graduating from these colleges, a college specialising in the production of science and mathematics teachers has recently been established.

Over the past two decades, many initiatives have been launched in science and mathematics education by non-government organisations. Many of these have focused on in-service training, but have been forced to close through the shortage of funding since the elections. The Department of Education will have to resume this part of the training on a larger scale to take the place of these organisations and to retrain teachers for the new curriculum. Alternatively, it will have to seek ways to delegate some of this responsibility to the still-existing organisations and strive for closer collaboration in this area.

7.6   Information and communications technology and science and mathematics education

There are clear indications that a linear expansion of South Africa's current education and training systems is beyond the country's reach. An approach requiring transformation using interactive electronic technologies together with the most advanced telecommunications system and interactive multimedia courseware appears to be what is required. However, the challenge to South Africa is to seek and select appropriate combinations of electronic media, face-to-face delivery and 'learning at a distance' delivery approaches, otherwise known as a flexible learning system. Different technologies have different suitabilities and have to be assessed accordingly for their learning gains, appropriateness, quality, efficiency and ultimately, their cost-effectiveness. The challenge in the future will be to cater for mass audience needs (especially in the informal and non-formal education sectors) and the more specialist needs-driven course in the formal sector.

7.7   Partnerships to reform science and mathematics education

Partnerships will play an important role in the future of mathematics and science education in South Africa. These will be between the Department of Education, Department of Trade and Industry, Department of Labour, the Department of Arts Culture, Science and Technology, industry and the private sector, schools, communities, parents, schools and their local business community. Local communities and local business must become more involved if the financial challenges facing mathematics and science education (in particular) are to be solved. Parents will have to pay school fees if their children are to benefit from their education. Attitudes of entitlement (to free education) will have to be overcome and the culture of teaching and learning restored before mathematics and science education can make headway in these challenging times. Strict financial planning and management will need to be implemented in order to allocate the resources needed for mathematics and science textbooks, science equipment and facilities for underprivileged schools. With the devolution of power to the provinces as well as financial control over spending, provincial authorities will have the responsibility for planning and addressing their local needs. The active promotion of science education for sustainable development by all must be undertaken by national and provincial government, commerce and industry and by communities, together.

Collaboration between the four principal government departments mentioned above is essential to ensure the links between science and technology policies with science education policy, ultimately leading to the successful reform of science education in South Africa.

8.   FINAL WORD

In Africa, most countries after independence placed the bulk of their efforts into curriculum development in school science, mathematics and technology.   However, due to several factors such as instability in government, growing populations, unemployment and collapsed economies, these countries have been unable to achieve the goal of self-reliance with respect to scientific and technological human resource development. Two of the primary reasons for the failure have been a lack of emphasis in the policies on science, technology and mathematics education as a route to science and technology human resource development and, secondly, inadequate statistics on which to base development plans.   South Africa needs to take cognisance of this experience in other African countries and the lessons from both developing and developed countries around the world. International studies offer South Africa the possibility of doing this by providing the necessary benchmarks and building local expertise to assist with development plans in education.

If South Africa is going to compete effectively in world markets, a well-educated corps of people with knowledge and skills at all levels in mathematics, science, technology, economics and communication will be needed.   Government leadership will be crucial, as it will have to demonstrate commitment and political will for the sake of long-term benefit to the whole country, to drive this process forward publicly.

To produce this effective, well-educated workforce, well-qualified and motivated teachers of these subjects are needed. The reason most often stated by pupils going into careers in science and technology is the quality of the mathematics and science teachers at school. Only with competent, dedicated, inspiring teachers will reform in science and mathematics education be possible.

Monitoring government plans by way of involvement in international studies is one way of objectively assessing positive (or negative) trends in education.   The information gathered by TIMSS put the spotlight on mathematics and science education in South Africa and contributed to the introduction of various initiatives in mathematics and science education in this country. Comparing South Africa's performance and education system with other countries' education systems may result in many lessons being learnt for the future and thereby the successful implementation of science and mathematics education reform in South Africa.

REFERENCES

African National Congress (1994) *A Policy Framework For Education and Training*, Education Department, ANC, p.83.

Beaton, A. E., Mullis, I.V.A., Martin, M., Gonzalez, E., Kelly, D. L., Smith, T. A. (1996a) *Third International Mathematics and Science Study, Mathematics Achievement in the Middle School Years* (Chestnut Hill, MA: Boston College)

Beaton, A. E., Martin, M., Mullis, I. V. A., Gonzalez, E., Smith, T. A. Kelly, D. L. (1996b), *Third International Mathematics and Science Study, Science Achievement in the Middle School Years* (Chestnut Hill, MA: Boston College)

Blankley, W. (1994) The abyss in African school education in South Africa. *South African Journal of Science*, Vol. 90, (02/94). p. 54

Brickman, W. W. (1988) History of comparative education, in Postlethwaite, T.N. (ed.) *The Encyclopedia of comparative education and national systems of education* (New York: Pergamon Press)

Bot, M. (1998) Education spending in the provinces. *Edusource Data News*. No.20. March. (Johannesburg: Education Foundation)

de Villiers, A. P. (1997) Inefficiency and demographic realities of the South African school system, *South African Journal of Education*, 17 (2) pp.79-80

Finn, C. E. (1989) A world of assessment, a universe of data, in Purves, A. (ed.) *International comparisons and educational reform* (Association for Supervision and Curriculum Development, USA)

FRD (Foundation for Research Development) (1993) SA Science and Technology Indicators, Pretoria

FRD (Foundation for Research Development (1996) SA Science and Technology Indicators, Pretoria, pp. 2-18

Gouws, E. (1997) Entrepeneurship education: an educational perspective, *SA Journal of Education*, Vol. 17 (3), pp. 143.

Greaney, V. and Kellaghan, T. (1996) *Monitoring the learning outcomes of educational systems* (Washington, DC: World Bank)

Heyneman, S. P. (1986) *The Search for school effects in developing countries: 1966-1986*, Seminar paper series no 33 (Washington, DC: The World Bank)

Howie, S. J., (1997) *Mathematics and science performance in the middle school years in South Africa: A summary report on the performance of the South Africa students in the Third International Mathematics and Science Study*. (HSRC published report, Pretoria)

Howie, S. J. (1998a) *Challenges facing the reform of science education in South Africa: What do the TIMSS results mean for South Africa?* In World Bank Series on Science Education in Developing Countries (Washington, DC: World Bank)

Howie, S. J., (1998b) *Producing quality science teachers: Challenges to Teacher Education*, *International Society for Teacher Education*, Skukuza, South Africa, April

Howie, S.J, and Hughes, CA. (1998) *Mathematics and science literacy of final-year school students in South Africa: Third International Mathematics and Science Study* (HSRC published report, Pretoria)

Human Sciences Research Council (1997) *Schools Needs Based Survey Database* (HSRC, Pretoria)

Howie, S. and Wedepohl, P. (1993) *Investigation of science teacher education in five sub-Saharan countries – a comparative overview* (FRD report, Pretoria)

Interpellations, Questions and replies of the National Assembly, (1997) Wednesday, 19 February, pp. 111-114

Interpellations, Questions and replies of the National Assembly, (1997a) Wednesday, 27 August, pp. 2258-2261

Keeves, J. P. (1992) *Learning science in a changing world. Cross-national studies of science achievement: 1970-1984.* International Association for the Evaluation of Educational Achievement. (The Hague)

Lockheed, M. (1988) *The measurement of educational efficiency and effectiveness. Paper presented at American Educational Research Association Annual Meeting,* (New Orleans)

Mullis, I. V. A., Martin, M., Beaton, A. E., Gonzalez, E., Kelly, D. L., Smith, T. A. (1998) *Mathematics and Science Achievement in the Final year of Secondary School* (Chestnut Hill, MA: Boston College)

Postlethwaite, T.N. (1988) *Cross national convergence of concepts and measurement of educational achievement,* a paper presented at American Educational Research Association Annual Meeting (New Orleans)

Rotberg, I.C. (1991) Myths in international comparisons of science and mathematics achievement. *The Bridge,* vol. 21, pp. 3-10

van Eldik P. (1998) *The role of higher education in technological innovation in economically developing countries,* a paper presented at the 38th Annual forum of the Association for Institutional research, May 17-20, (Minneapolis)

White Paper on Science and Technology (1996), *Preparing for the 21st Century.* (Department of Arts, Culture, Science and Technology, Pretoria)

ACKNOWLEDGEMENTS

My thanks to the World Bank for its sponsorship and for making South Africa's participation in TIMSS and TIMSS-R possible, and to the University of Leeds for affording me the opportunity to speak at the 1998 conference. Finally, my sincere thanks to Professor Peter van Eldik who contributed significantly to a previous draft of this chapter.

BIBLIOGRAPHY

Adams R. J. and Gonzalez E.J. (1996) The TIMSS Test Design, in Martin, M. O and Kelly, D. L. *The Third International Mathematics and Science Study, Technical Report, Volume 1: Design and Development* (Chestnut Hill, MA: Boston College)

African National Congress (1994) *A Policy Framework For Education and Training,* (Education Department, ANC)

Aitkin, M. and Longford, N. (1986) Statistical modeling issues in school effectiveness studies, *Journal of the Royal Statistical Society, Series A*, 149 (1), pp. 1-43

Anderson, B. (1986) *Homework: what do national assessment results tell us?* (Princeton, N.J., Educational Testing Service)

Angell, C. (1995) *Codes for Population 3, Physics Specialists, Free Response Items.* TIMSS report no. 16 (University of Oslo)

Angell, C. and Kobberstad, T. (1993) *Coding Rubrics for Free-Response Items.* (Doc.Ref.: ICC800/NRC360). Paper prepared for the Third International Mathematics and Science Study (TIMSS)

Angell, C., Brekke, G., Gjørtz, T., Kjærnsli, M., Kobberstad, T., and Lie, S. (1994) *Experience with Coding Rubrics for Free-Response Items.* (Doc.Ref. ICC867). Paper prepared for the Third International Mathematics and Science Study (TIMSS)

Ashton, D. N. and Maguire, M. (1986) Young Adults in the Labour Market, *Research Paper 55* (London: Department of Employment)

Assor, A. and Connell, J. P. (1992) The validity of students' self-reports as measures of performance affecting self-appraisals, in: D. H. Schunk and J. L. Meece (eds.) *Student Perceptions in the Classroom* (Hillsdale, New Jersey, Lawrence Erlbaum)

Atkin, J. M. and Black, P. J. (1997) Policy perils of international comparisons: the TIMSS case, *Phi Delta Kappan*, pp. 22-28 (September)

Baker, P. D. (1997) Good news, bad news, and international comparisons: Comment on Bracey, *Educational Researcher, 26* (3), pp. 16-17

Barber, M. (1996) *The Learning Game* (London, Victor Gollancz)

Bates, I. (1995) The Competence Movement: Conceptualising Recent Research, *Studies in Science Education*, 25, pp. 39-68

Beaton, A. E., Martin, M. O., Mullis, I. V. S., Gonzalez, E. J., Smith, T. A., and Kelly, D. L. (1996) *Science Achievement in the Middle School Years: IEA's Third International Mathematics and Science Study (TIMSS)* (Chestnut Hill, MA: Boston College)

Beaton, A. E., Mullis, I. V. S, Martin, M. O., Gonzalez, E.J., Kelly, D. L. and Smith, T. A. (1996) *Mathematics Achievement in the Middle School Years: I.E.A.'s Third International Mathematics and Science Study (TIMSS)* (Chestnut Hill, MA: Boston College)

Beaton, A. E. (1996) Preface, in Robitaille, D.F. and Garden, R. A. *Research Questions and Study Design*, TIMSS Monograph No. 2, (Vancouver: Pacific Educational Press)

Ben-Peretz, M., and Zeidman, A. (1986) Three generations of curricular development in Israel. *Studies in Education, 43/44,* pp. 317-327

Bishop, J. H. (1989) Why the apathy in American high schools? *Educational Researcher*, 18, pp.7-10

Blankley, W. (1994) The abyss in African school education in South Africa. *South African Journal of Science*, Vol. 90, (02/94) p.54

Blatchford, P. (1996) Pupils' views on school work and school from 7 to 16 years, *Research Papers in Education*, 11(3), pp. 263-288

Boekaerts, M. (1994) *Motivation in Education: the 14th Vernon-Wall Lecture* (Leicester: British Psychological Society)

Bot, M. (1998) Education spending in the provinces. *Edusource Data News.* Vol. 20. March. (Johannesburg: Education Foundation)

Bracey, G. W. (1996) International comparisons and conditions of American education. *Educational Researcher, 25* (1), pp. 5-11

Bracey, G. W. (1997) On comparing the incomparable: A response to Baker and Stedman. *Educational Researcher, 26* (3), pp. 19-25

Brickman, W. W. (1988) History of comparative education, in Postlethwaite, T.N. (ed.) *The Encyclopedia of comparative education and national systems of education* (New York: Pergamon Press)

Bridgman, A (1994) Teaching the old bear new tricks, *The American School Board Journal,* 181(2), pp. 26-31

Bronfenbrenner, U. (1971) *The Two Worlds of Childhood* (London: Allen and Unwin)

Burghes, D. (1996) *Education across the world: the Kassel Project.* Paper presented at conference 'Lessons from Abroad: Learning from international experience in education' (London)

Burton, D. (1997) The myth of 'expertness': cultural and pedagogical obstacles to restructuring East European curricula, *British Journal of In-service Education*, 23(2), pp. 219-229

Chen, D., and Novick, R. (1987) The challenge is scientific and technological literacy for all. *MABAT Publication, 5* (School of Education, Tel Aviv University [Hebrew])

Chen, M., Lewy, A., and Adler, H. (1978) *Processes and outcomes of educational practice: Assessing the impact of junior high school* (School of Education, Tel Aviv University and the NCJW Research Institute for Innovation in Education, School of Education, The Hebrew University, Jerusalem [Hebrew])

Chuprov, V. and Zubok, I. (1997) Social conflict in the sphere of the education of youth. *Education in Russia, the Independent States and Eastern Europe,* 15(2), pp. 47-58

Coleman, J. (1961) *Adolescent Society: The Social Life of the Teenager and its Impact on Education* (New York: Free Press)

Cooper, H. (1989) *Homework* (White Plains, N.Y.: Longman).

Cooper, H., Lindsey, J. L., Nye, B. and Greathouse, S. (1998) Relationships among attitudes about homework, amount of homework assigned and completed, and student achievement, *Journal of Educational Psychology*, 90(1), pp. 70-83

Covington, M. V. (1992*) Making the Grade: A Self-Worth Perspective on Motivation and School Reform* (Cambridge: Cambridge University Press)

Crandall, V. C., Katkovsky, W. and Crandall, V. J. (1965) Children's beliefs in their own control of reinforcements in intellectual-academic situations. *Child Development*, 36, pp. 91-109

Creemers, B. (1992) School effectiveness and effective instruction – the need for a further relationship. In Bashi, J. and Sass, Z. (eds.), *School Effectiveness and Improvement*, (Jerusalem: Hebrew University Press)

Creemers, B. P. M. (1994) *The effective classroom* (London: Cassell)

Creemers, B. P. M. and Reynolds, D. (1996) Issues and Implications of International Effectiveness Research, *International Journal of Educational Research*, 25 (3), pp. 257-266.

Creemers, B. P. M. and Scheerens, J. (eds.) (1989) Developments in school effectiveness research. A special issue of *International Journal of Educational Research*, 13 (7), pp. 685-825

Dan, Y. (1983) The process of becoming a teacher education college. *Mahalachim: The Levinsky Teachers College Annual*, pp. 22-32 (Hebrew)

de Villiers, A. P. (1997) Inefficiency and demographic realities of the South African school system, *South African Journal of Education*, 17 (2) pp.79-80

di Sessa, A. A. (1993) Toward an Epistemology of Physics. *Cognition and Instruction*, 10 (2 and 3), pp. 105-225

Driver, R. and Easley, J. (1978) Pupils and Paradigms: A Review Literature Related to Concept Development in Adolescent Science Students. *Studies in Science Education*, 5, pp. 61-83

Dweck, C. (1986) Motivational processes affecting learning, *American Psychologist*, 41, pp. 1040-1048

Eaton, M.J. and Dembo, M.H. (1997) Differences in the motivational beliefs of Asian American and Non-Asian students, *Journal of Educational Psychology*, 89 (3), pp. 433-440

Ebison, M. G. (1993) Newtonian in Mind but Aristotelian at Heart. *Science and Education* 2, pp. 345-362

Elliott, J., Hufton, N., Illushin, L. and Lauchlan, F. Motivation in the junior years: international perspectives on children's attitudes, expectations and behaviour and their relationship to educatinal achievement (submitted)

Ellis, D. (1997) Maths study shows cause for concern. *The Adelaidian'* November, pp 1-3 (University of Adelaide)

Epstein, L.L. (1988) *Homework Practices, Achievements and Behaviors of Elementary School Students* (Baltimore MD, Center for Research on Elementary and Middle Schools)

Etzioni Committee (1979) *The Report of the Committee Appointed to Assess the Status of Teachers and Teaching* (Jerusalem, Ministry of Education and Culture. Also published in *Hed Ha'chinuch, 55* (18), pp. 7-22 [Hebrew])

Farrow, S., Tymms, P. and Henderson, B. (1999) Homework and attainment in primary schools, *British Educational Research Journal*, 25(3), pp. 323-341

Finegold, M. and Gorsky, P. (1991) Students' concept of force as applied to related physical systems: A search for consistency, *International Journal of Science Education*, 13, 1, pp. 97-113

Finn, C. E. (1989) A world of assessment, a universe of data, in Purves, A. (ed.) *International comparisons and educational reform* (Association for Supervision and Curriculum Development, USA)

Forgione, P. D. (1998) *What We've Learned From TIMSS About Science Education in the United States,* Address to the 1998 Conference of the National Science Teachers' Association (http://nces.ed.gov/timss/report/97255-04.html)

Foxman, D. (1992) *Learning Mathematics and Science, The Second International Assessment of Educational Progress in England)* (Slough: NFER)

FRD (Foundation for Research Development (1996) SA Science and Technology Indicators (Pretoria, pp. 2-18)

FRD (Foundation for Research Development) (1993) SA Science and Technology Indicators (Pretoria)

Freudenthal, H. (1975) Pupils' achievement internationally compared – the IEA, *Educational Studies in mathematics*, 6, pp. 127-186

Garden, R. A. and Orpwood, G. (1996) Development of the TIMSS Achievement Tests, in Martin, M.O and Kelly, D. L. *The Third International Mathematics and Science Study, Technical Report, Volume 1: Design and Development* (Chestnut Hill, MA: Boston College)

Gipps, C. (1996) The paradox of the Pacific Rim learner, *Times Educational Supplement*, 20th December, p. 13.

Gipps, C. and Tunstall, P. (1998) Effort, ability and the teacher: young children's explanations for success and failure, *Oxford Review of Education*, 24 (2), pp. 149-165

Glowka, D. (1995) *Schulen und unterricht im vergleich. Rusland/Deutschland (Schools and teaching in comparison: Russia and Germany)* (New York: Waxmann Verlag)

Goldstein, H. (1996) Introduction to general issues. The IEA Studies Assessment in Education, *Principal Policy and Practice, 3* (2), pp. 125-128

Goldstein, H. (1995) *Multilevel models in educational and social research : A Revised Edition* (London: Edward Arnold)

Goldstein, H. (1996) International Comparisons of student achievement, in Little, A. and Wolfe, A. (eds.), *Assessment in Transition: learning, monitoring and selection in international perspective* (London: Pergamon)

Gouws, E. (1997) Entreppeneurship education: an educational perspective, *SA Journal of Education,* Vol 17 (3), p. 143

Greaney, V. and Kellaghan, T. (1996) Monitoring the learning outcomes of educational systems (Washington, D.C: World Bank)

Griffith, J. E., and Medrich, E. A. (1992) What does the United States want to learn from international comparative studies in education?, *Prospects*, Vol. 22, pp. 476-485

Han, Jong-Ha. (1995) The Quest for National Standards in Science Education in Korea, *Studies in Science Education*, 26, pp. 59-71

Hans, N. *Comparative Education: A Study of Educational Factors and Tradition* (London)

Harmon, M., Smith, T. A., Martin, M. O., Kelly, D. L., Beaton, A. E., Mullis, I. V. S., Gonzalez, E . J., and Orpwood, G. (1997) *Performance Assessment in IEA's Third International Mathematics and Science Study* (Chestnut Hill, MA: Boston College)

Harris, S., Keys, W. and Fernandes, C. (1997) *Third International Mathematics and Science Study: Second National Report, Part 1* (Slough: NFER)

Heyneman, S. P. (1986) The Search for school effects in developing countries: 1966-1986, *Seminar paper series no 33* (Washington, DC: The World Bank)

Hirschman, C. and Wong, M. G. (1986) The extraordinary educational attainment of Asian Americans: A search for historical evidence and explanations, *Social Forces*, 65, pp. 1-27

Howie, S. and Wedepohl, P. (1993) *Investigation of science teacher education in five sub-Saharan countries – a comparative overview* (Pretoria: FRD)

Howie, S. J. (1997) *Mathematics and science performance in the middle school years in South Africa: A summary report on the performance of the South Africa students in the Third International Mathematics and Science Study* (Pretoria: HSRC)

Howie, S. J. and Hughes, C. A. (1998) *Mathematics and science literacy of final-year school students in South Africa: Third International Mathematics and Science Study* (Pretoria: HSRC)

Howie, S. J. (1998a) Challenges facing the reform of science education in South Africa: What do the TIMSS results mean for South Africa?, in *World Bank Series on Science Education in Developing Countries* (Washington, DC: World Bank)

Howie, S. J., (1998b) Producing quality science teachers: Challenges to Teacher Education, *International Society for Teacher Education*, April (Skukuza: South Africa)

Howley, C. B., Howley, A. and Pendarvis, E. D. (1995) *Out of Our Minds: Anti-Intellectualism and Talent Development in American Schooling* (New York: Teachers College Press)

Howson, G. (1998) The Value of Comparative Studies, in Kaiser, G., Luna, E., and Huntley, I (Eds) *International Comparisons in Mathematics Education* (London: Falmer Press)

Hufton, N. and Elliott, J. (2000) Motivation to learn: the pedagogical nexus in the Russian school: some implications for transnational research and policy borrowing, *Educational Studies*, 26, pp. 115-136

Human Sciences Research Council (1997) *Schools Needs Based Survey Database* (Pretoria: HSRC)

Hussein, M. G. (1992) What does Kuwait want to learn from the Third International Mathematics and Science Study (TIMSS)?, *Prospects*, 22, pp. 63-468

Interpellations, Questions and replies of the National Assembly (1997) Wednesday, 19 February, pp. 111-114

Interpellations, Questions and replies of the National Assembly (1997a) Wednesday, 27 August, pp. 2258-2261

Israel Believes in Education (1996) (Jerusalem: Ministry of Education, Culture and Sports)

Kaiser, G., Luna, E. and Huntly. I. (eds.) *International Comparisons in Mathematics Education* (London: Falmer)

Kaiser, G. (1999) Comparative Studies on Teaching Mathematics in England and Germany. In Kaiser, G., Luna, E. and Huntly, L. (eds.): *op.cit.,* pp. 140-50

Kandel, I.L. (1955) *The New Era in Education: A Comparative Study* (London: Harrap)

Karakovsky, V.A. (1993) The school in Russia today and tomorrow, *Compare*, 23 (3), pp. 277-288

Keeves, J. P. (1992). *Learning science in a changing world: Cross-national Studies of Science Achievement: 1970-1984* (The Hague: IEA)

Keitel, C. and Kilpatrick, J. (1999) Rationality and Irrationality of International Comparisons. In G. Kaiser, E. Luna and I. Huntly (eds.), *International Comparisons in Mathematics Education,* pp. 241-56 (London: Falmer)

Keith, T.Z. (1986) Children and homework, in: A. Thomas and J. Grimes (eds.) *Children's needs: psychological perspectives* (Washington, DC: NASP)

Keys, W (1999) *What can mathematics educators in England learn from TIMSS? Educational Research and Evaluation*, 5, pp. 195-213

Keys, W. and Fernandes, C. (1993) *What do students think about school?* (Slough: NFER)

Keys, W., Harris, S. and Fernandes, C. (1996) *Third International Mathematics and Science Study, First National Report, Part 1* (Slough: NFER)

Keys, W., Harris, S. and Fernandes, C. (1997) *Third International Mathematics and Science Study, First National Report, Part 2* (Slough: NFER)

Kitaev, I. V. (1994) Russian education in transition: transformation of labour market, attitudes of youth and changes in management of higher and lifelong education, *Oxford Review of Education,* 20 (1), pp. 111-130

Kjaernsli, M., Kobberstad, T., and Lie, S. (1994) *Draft Free-Response Coding Rubrics-Populations 1 and 2* (Doc.Ref: ICC864) Document prepared for the Third International Mathematics and Science Study (TIMSS)

Levine, D. U. and Lezotte, L. W. (1990) *Unusually effective schools : A review and analysis of research and practice.* (Madison, WI, The National Center for Effective Schools Research and Development)

Lewy, A., and Chen, M. (1974) *Closing or widening of the achievement gap: A comparison over time of ethnic group achievement in the Israel elementary school.* Research Report No. 6, (School of Education, Tel Aviv University [Hebrew])

Lie, S., Taylor, A., and Harmon, M. (1996) *Scoring Techniques and Criteria*. Chapter 7 in Martin, M.O. and Kelly, D. L. (eds.): Third International Mathematics and Science Study, Technical Report, Volume 1: Design and Development. (Chestnut Hill, MA: Boston College)

Lightbody, P. Siann, G., Stocks, R and Walsh, D. (1996) Motivation and attribution at
secondary school: the role of gender, *Educational Studies,* 22 (1), pp. 13-25

Lokan, J., Ford, P. & Greenwood, L. (1996) *Maths and Science on the Line: Australian
Junior Secondary Students' Performance in the Third International Mathematics and
Science Study.* TIMSS Australia Monograph No. 1 (Melbourne: Australian Council for
Educational Research)

Lokan, J. & Ainley, J. (1998) The Third International Mathematics and Science Study:
Implications for the development of numeracy benchmarks. *Unicorn,* 24, pp. 97-109.

Lockheed, M. (1988) The measurement of educational efficiency and effectiveness. Paper
presented at American Educational Research Association Annual Meeting (New Orleans)

Loxley, W (1992) Introduction to special volume, *Prospects,* 22 pp. 275-277

McGaw, B., Long, M., Morgan, G. & Rosier, M. (1989) *Literacy and Numeracy in Victorian
Schools: 1988.* ACER Research Monograph No 34 (Melbourne: Australian Council for
Educational Research)

McInerney, D. M., Roche, L. A., McInerney, V. and Marsh, H. (1997) Cultural perspectives
on school motivation: the relevance and application of goal theory, *American Educational
Research Journal,* 34(1), pp. 207-236

McLean, K. (1997) Towards a National Report Card. *Education Quarterly,* No. 3, Spring,
pp. 5-7

Martin, M. O., Mullis, I. V. S., Beaton, A. E., Gonzalez, E. J., Smith, T. A., and Kelly, D. L.
(1997) *Science Achievement in the Primary School Years: IEA's Third International
Mathematics and Science Study (TIMSS)* (Chestnut Hill, MA: Boston College)

Martin, M. O. and Kelly, D. L., (eds.) (1996) *TIMSS Technical Report, Volume I: Design and
Development.* (Chestnut Hill, MA: Boston College)

Martin, M. O. and Kelly, D. L., (eds.) (1997) *TIMSS Technical Report, Volume II:
Implementation and Analysis, Primary and Middle School Years* (Chestnut Hill, MA:
Boston College)

Martin, M. O. and Kelly, D. L., (eds.) (in press) *TIMSS Technical Report, Volume III:
Implementation and Analysis, Final Year of Secondary School* (Chestnut Hill, MA: Boston
College)

Martin, M. O. and Mullis, I. V. S. (eds.) (1996) *Quality Assurance in Data Collection*
(Chestnut Hill, MA: Boston College)

Martin, M. O., Mullis I. V. S., Beaton, A. E., Gonzales, E. J., Smith, T. A., and Kelly, D. L.
(1997): *Science achievement in the Primary School Years. IEA's Third International
Mathematics and Science Study (TIMSS)* (Chestnut Hill, MA: Boston College)

Morris, P. (1998) Comparative education and educational reform: beware of prophets
returning from the Far East, *Education,* 3-13, pp. 3-7

Muijs, R. D. (1997) Predictors of academic achievement and academic self-concept: a
longitudinal perspective, *British Journal of Educational Psychology,* 67 (3), pp. 263-277

Mullis I. V. S., Martin, M. O., Beaton, A. E., Gonzales, E.J., Kelly, D. L., and Smith, T. A.
(1998) *Mathematics and Science Achievement in the Final Year of Secondary School.
IEA's Third International Mathematics and Science Study (TIMSS).* (Chestnut Hill, MA:
Boston College)

Mullis, I. V .S., Martin, M. O., Beaton, A. E., Gonzalez, E. J., Kelly, D. L., and Smith, T. A.
(1997). *Mathematics Achievement in the Primary School Years: IEA's Third International
Mathematics and Science Study (TIMSS).* (Chestnut Hill, MA: Boston College)

Mullis, I. V. S. and Smith, T. A. (1996): *Quality Control Steps for Free-Response Scoring.*
Chapter 5 in Martin, M. O. and Mullis I. V. S. (eds.) Third International Mathematics and
Science Study: Quality Assurance in Data Collection (Chestnut Hill, MA: Boston College)

Nebres, B. F. (1999) *International Benchmarking as a Way to Improve School Mathematics Achievement in the Era of Globalisation*, in G. Kaiser, E. Luna and I. Huntly (eds.), *op.cit.*, pp. 200-12

Nesbit, J. (1974) Fifty Years of Research in Science Education, *Studies in Science Education*, 1, pp. 103-12

Nikandrov, N. D. (1995) Russian education after perestroika: the search for new values, *International Review of Education,* 41(1-2), pp. 47-57

Noah, H. J. and Eckstein, M. A. (1969) *Towards a Science of Comparative Education* (Toronto: Macmillan)

Oettingen, G., Little, T. D., Lindenberger, U. and Baltes, P. B. (1994) Causality, agency and control beliefs in East versus West Berlin children: a natural experiment on the role of context, *Journal of Personality and Social Psychology*, 66, pp. 579-595

Office for Standards in Education (1993) *Access and Achievement in Urban Education* (London: HMSO)

Ogbu, J. (1983) Minority status and schooling in plural societies, *Comparative Education Review*, 27, pp. 168-190

Olson, L. (1994) Dramatic rise in Kentucky test scores linked to reforms. *Education Week*, 13, p.13

Olympia, D. E., Sheridan, S. M. and Jenson, W. (1994) Homework: a natural means of home-school collaboration, *School Psychology Quarterly,* 9(1), pp. 60-80

Orpwood, G. (ed.) (1998) *Ontario in TIMSS: Secondary Analysis and Recommendations*, Unpublished report prepared on behalf of the Ontario Association of Deans of Education for the (Ontario) Educational Quality and Accountability Office

Orpwood, G. and Garden, R. A. (1998) *Assessing Mathematics and Science Literacy.* TIMSS Monograph No 4, (Vancouver: Pacific Educational Press)

Osborn, M. (1997) Children's experience of schooling in England and France: some lessons from a comparative study, *Education Review*, 11(1), pp. 46-52

Peled, A. (1976) *The education in Israel during the 80s.* (Jerusalem, Ministry of Education and Culture [Hebrew])

Pfundt, H and Duit, R. (1994) *Students' Alternative Frameworks and Science Education.* 4th Edition (Germany: IPN at the University of Kiel)

Politcia, *Comparing Standards (London: Politeia Press)*

Postlethwaite, T. N. (1988) *The Encyclopaedia of Comparative Education and National Systems of Education, Preface* (Oxford: Pergamon Press)

Postlethwaite, T. N. (1988) *Cross national convergence of concepts and measurement of educational achievement.* A paper presented at American Educational Research Association Annual Meeting (New Orleans)

Poznova, G. (1998) Russian teenagers: what do they strive for today? *Education in Russia, the Independent States and Eastern Europe*, 19 (1), pp. 19-25

Prais, S. (1997) *School-readiness, whole class teaching and pupils' mathematical attainments* (London: National Institute of Economic and Social Research)

Prelovskaya, I. (1994) New problems in the market economy, *Izvestia*, 26th June, pp. 1-2

Purves, A. C. (1992) *The IEA Study of Written Composition II : Education and Performance in Fourteen Countries* (Oxford: Pergamon Press)

Reynolds, D. (1997) Good ideas can wither in another culture, *Times Education Supplement*, 19th September, p.2

Reynolds, D. and Farrell, S. (1996) *Worlds Apart? – A Review of International Studies of Educational Achievement Involving England.* (London: HMSO for OFSTED)

Reynolds, D., Creemers, B. P. M., Stringfield, S., Teddlie, C., Schaffer, E. and Nesselrodt, P. (1994) *Advances in School Effectiveness Research and Practice* (Oxford: Pergamon Press)

Robinson, P. (1997) *Literacy, Numeracy and Economic Performance* (London: Centre for Economic Performance)

Robitaille, D.F. (ed.) (1997) *National Contexts for Mathematics and Science Education: An Encyclopaedia of the Education Systems Participating in TIMSS* (Vancouver: Canada, Pacific Educational Press)

Robitaille, D.F. (ed.) (1997) *National Contexts for Mathematics and Science Education. An Encyclopaedia of the Education Systems Participating in TIMSS* (Vancouver: Pacific Educational Press)

Robitaille, D. F. and Garden, R. A. (1996) *Research Questions and Study Design.* TIMSS Monograph No. 2 (Vancouver: Pacific Educational Press)

Robitaille, D. F. and Robeck, E. C. (1996) The Character and the Context of TIMSS, in Robitaille, D.F. and Garden, R.A. (eds). *Research Questions and Study Design,* TIMSS Monograph No 2 (Vancouver: Pacific Educational Press)

Robitaille, D. F., Schmidt, W. H., Raizen, S. A., McKnight, C. C., Britton, E. D., and Nicol, C. (1993) *Curriculum Frameworks for Mathematics and Science.* TIMSS Monograph No. 1 (Vancouver: Pacific Educational Press)

Rosenbaum, J. E. (1989) What if good jobs depended on good grades? *American Educator* (Winter), pp. 1-15, pp. 40-42

Rosier, M. J. (1980) *Changes in Secondary School Mathematics in Australia: 1964–1978.* ACER Research Monograph No. 8 (Melbourne: Australian Council for Educational Research)

Ross, K. N. (1992) Sample design for international studies, *Prospects,* 22 pp. 305-315

Rotberg, I. C. (1991) Myths in international comparisons of science and mathematics achievement, *The Bridge,* Vol. 21, pp. 3-10

Sarigiani, P. A., Wilson, J. L., Petersen, A. C. and Vicary, J. R. (1990) Self-image and educational plans of adolescents from two contrasting communities, *Journal of Early Adolescence,* 10 (1), pp. 37-55

Schaffer, E. (1994) The contributions of classroom observation to school effectiveness research, in Reynolds, D., Creemers, B. P. M, Nesselrodt, P., Schaffer, E., Stringfield, S. and Teddlie, C. *Advances in School Effectiveness Research and Practice.* (Oxford: Pergamon)

Scheerens, J. (1992) *Effective schooling : Research, theory and practice.* (London: Cassell)

Scheerens, J. and Bosker, R. (1997) *The Foundations of School Effectiveness* (Oxford: Pergamon Press)

Schleicher, A. and Umar, J. (1992) Data management in educational survey research, *Prospects* 22, pp. 318-325

Schmida, M. (1987) *Equality and excellence: Educational reform and the comprehensive school.* (Ramat Gan: Bar-Ilan, University Press [Hebrew])

Schmidt, W. H, Raizen, S. A., Britton, E. D., Bianchi, L. J, and Wolfe, R. G. (1997) *Many Visions, Many Aims, Volume 2: A Cross-National Investigation of Curricular Intentions in Science* (Dordrecht: Kluwer)

Schmidt, W. H. and McKnight, C. C. (1998) *So What Can We Really Learn from TIMSS?* Paper published on the Internet (http://TIMSS.msu.edu/whatcanwelearn.htm)

Schmidt, W. H., Jorde, D., Cogan, L. S., Barrier, E., Gonzalo, I., Moser, U., Shimizu, Y., Sawada, T., Valverde, G. A., McKnight, C. C., Prawat, R., Wiley, D. E., Raizen, S. A., Britton, E. D., Wolfe, R. G. (1996) *Characterising Pedagogical Flow: An Investigation of Mathematics and Science teaching in Six Countries* (Dordrecht: Kluwer)

Schmidt, W. H., McKnight, C. C., Valverde, G. A., Houang, R. T., and Wiley, D. E. (1997a) *Many Visions, Many Aims: A Cross-National Investigation of Curricular Intentions in School Mathematics* (Dordrecht: Kluwer)

Schmidt, W. H., Raizen S. A., Britton, E. D., Bianchi, L. J ., and Wolfe, R. G. (1997b) *Many Visions, Many Aims: A Cross-National Investigation of Curricular Intentions in School Science* (Dordrecht: Kluwer)

Sedlak, M., Wheeler, C. W., Pullin, D. C. and Cusick, P. A. (1986) *Selling students short: Classroom bargains and academic reform in the American high school* (New York: Teachers College Press)

Shavelson, R., McDonnell, L., Oakes, J. (eds.) (1989) *Indicators for monitoring mathematics and science education: A sourcebook* (Santa Monica, CA: Rand)

Shorrocks-Taylor, D and Hargreaves, M (1999) Making it clear: a review of language issues in testing, with special reference to the National Curriculum mathematics tests at Key Stage 2. *Educational Research,* Vol 41, No 2, pp. 123-136

Shorrocks-Taylor, D., Jenkins, E., Curry, J., Swinnerton, B., Hargreaves, M. and Nelson, N. (1998) *An Investigation of the Performance of English Pupils in the Third International Mathematics and Science Study (TIMSS)* (Leeds: Assessment and Evaluation Unit/ Centre for Studies in Science and Mathematics Education, University of Leeds)

Simpson, S. M., Licht, B. G., Wagner, R. K. and Stader, S. R. (1996) Organisation of children's academic ability-related self-perceptions, *Journal of Educational Psychology,* 88 (3), pp. 387-396

Siu, S. F. (1992) *Toward an understanding of Chinese American educational achievement (Report No. 2)* (Washington DC: US Department of Health and Human Services, Center on Families, Communities, Schools and Children's Learning)

Sjøberg, S. and Lie, S. (1981) *Ideas about force and movement among Norwegian pupils and students,* (Report 81-11, University of Oslo)

Sjøberg, S. (1992) *The IEA Science Study, SISS. Some Critical Points on Items and Questionnaires*, paper presented at the Fifteenth Comparative Education Society in Europe Conference, Dijon, June 27th-July 2nd 1992

Spearritt, D. (1987) Educational Achievement. In Keeves, J. P. (ed.), *Australian Education: Review of Recent Research,* pp. 117-146 (Sydney: Allen & Unwin)

Stacey, K. (1999) Implications of TIMSS for Mathematics Education, in Lokan, J. (ed.), *Raising Australian Standards in Mathematics and Science: Insights from TIMSS.* Proceedings of ACER National Conference 1997 pp. 117-146 (Melbourne: Australian Council for Educational Research)

Stedman, L.C. (1997) International achievement differences. An assessment of new perspective. *Educational Researcher, 26* (3), pp. 4-17

Steinberg, L. (1996) *Beyond the Classroom: Why school reform has failed and what parents need to do* (New York: Touchstone Books)

Steinberg, L., Dornbusch, S. M. and Brown, B. B. (1992) Ethnic differences in adolescent achievement: An ecological perspective, *American Psychologist,* 47, pp. 723-729

Stevenson, H. W. (1992) Learning from Asian Schools. *Scientific American,* December, pp. 32-38

Stevenson, H. W. and LeTendre, G. (eds.) (1998) *The Educational System in Japan: Case Study Findings.* Publication No. SAI98-3008. (Washington, DC: US Department of Education, National Institute on Student Achievement, Curriculum, and Assessment)

Stevenson, H. W. and LeTendre, G. (eds.) (1999a)*The Educational System in Germany: Case Study Findings.* (Washington, DC: US Department of Education, National Institute on Student Achievement, Curriculum, and Assessment)

Stevenson, H. W. and LeTendre, G. (eds.) (1999b) *The Educational System in the United States: Case Study Findings* (Washington, DC: US Department of Education, National Institute on Student Achievement, Curriculum, and Assessment)

Stevenson, H. W. and Lee, S. (1990) A study of American, Chinese and Japanese Children, *Monographs of the Society for Research in Child Development,* No. 221, 55 pp. 1-2

Stevenson, H. W. and Stigler, J.W. (1992) *The Learning Gap: Why Our Schools Are Failing and What We Can Learn from Japanese and Chinese Education* (New York: Summit Books)

Stevenson, H. W., Chen, C. and Uttal, D. H. (1990) Beliefs and achievement: A study of Black, Anglo and Hispanic children, *Child Development,* 61, pp. 508-523

Stilger, J. W., Gonzales, P., Kawanaka, T., Knoll, S. and Serrano, A. (1998) *Methods and Findings of the TIMSS Videotape Classroom Study,* (Washington DC: US Government Printing Office)

Stipek, D. and Gralinski, J. H. (1996) Children's beliefs about intelligence and school performance, *Journal of Educational Psychology,* 88 (3), pp. 397-407

Teddlie, C. and Reynolds, D. (1999) *The International Handbook of School Effectiveness Research* (London: Falmer Press)

Third International Mathematics and Science Study (TIMSS) (1995a) *Coding Guide for Free-Response Items-Populations 1 and 2* (Doc.Ref.: ICC897/NRC433) (Chestnut Hill, MA: Boston College)

Third International Mathematics and Science Study (TIMSS) (1995b) *Coding Guide for Free-Response Items-Population 3* (Doc.Ref.: ICC913/NRC446) (Chestnut Hill, MA: Boston College)

TIMSS Mathematics Items, Released Set for Population 1 (Primary) (Chestnut Hill, MA: Boston College)

TIMSS Mathematics Items, Released Set for Population 2 (Middle) (Chestnut Hill, MA: Boston College)

TIMSS Released Item Set for the Final Year of Secondary School: Mathematics and Science Literacy, Advanced Mathematics, and Physics (Chestnut Hill, MA: Boston College)

TIMSS Science Items, Released Set for Population 1 (Primary) (Chestnut Hill, MA: Boston College)

TIMSS Science Items, Released Set for Population 2 (Middle) (Chestnut Hill, MA: Boston College)

Tomorrow 98 (1992) *The report of the high commission of scientific and technological education* (Israel Ministry of Education and Culture [Hebrew])

Tye, B.B. (1985) *Multiple realities: A study of 13 American high schools* (Lanham, MD: University Press of America)

US Department of Education, National Center for Educational Statistics (1998) *Pursuing Excellence: A Study of US Twelfth-Grade* (Washington DC: NCES, US Government Printing Office)

United States Department of Education, National Centre for Education Statistics (1996). *Pursuing Excellence: A Study of US Eighth-Grade Mathematics and Science Teaching, Learning, Curriculum, and Achievement in International* Context, (Washington DC: NCES, US Government Printing Office)

United States Department of Education, National Centre for Education Statistics (1997). *Pursuing Excellence: A Study of US Fourth-Grade Mathematics and Science Achievement in International Context,* (Washington DC: NCES, US Government Printing Office)

United States Department of Education, National Centre for Education Statistics (1998). *Pursuing Excellence: A Study of US Twelfth-Grade Mathematics and Science Achievement in International Context,* (Washington DC, NCES, US Government Printing Office)

United States Department of Education, Office of Educational Research and Improvement (1997). *Moderator's Guide to Eighth-grade Mathematics Lessons: United States, Japan, and Germany,* (Washington DC: NCES, US Government Printing Office) van de Grift, W. (1990) Educational leadership and academic achievement in secondary education. *School Effectiveness and School Improvement*, 1 (1), pp. 26-40

van Eldik P. (1998) *The role of higher education in technological innovation in economically developing countries*, a paper presented at the 38th Annual forum of the Association for Institutional research, May 17-20, Minneapolis

Viennot, L. (1979) Spontaneous Reasoning in Elementary Dynamics. *European Journal of Science Education*, 1, 2, no. 205-22

Vulliamy, G. (1987) School effectiveness research in Papua New Guinea, *Comparative Education*, 23 (2), pp. 209-223

Wainer, H. (1993) Measurement problems, *Journal of Educational Measurement*, 30, pp. 1-21

Wandersee, J. H., Mintzes J. J. and Novak, J. D. (1993) Research on alternative conceptions in science, in Gabel, D. (ed.) *Handbook of research on science teaching and learning*. (New York: Macmillan)

Watanabe, R. (1992) How Japan Makes Use of International Educational Survey research, *Prospects*, 22, pp. 456-62

Webber, S. K. (1996) Demand and supply: meeting the need for teachers in the 'new' Russian school, *Journal of Education for Teaching*, 22 (1), pp. 9-26

Weiner, B. (1979) A theory of motivation for some classroom experiences, *Journal of Educational Psychology*, 71, pp. 3-25

Whitburn, J. (2000) Strength in Numbers: Learning Maths in Japan and England (London: National Institute of Economic and Social Research)

White Paper on Science and Technology (1996) *Preparing for the 21st Century* (Pretoria: Department of Arts, Culture, Science and Technology)

Wilson, S., Henry, C. S. and Peterson, G. W. (1997) Life satisfaction among low-income rural youth from Appalachia, *Journal of Adolescence*, 20, pp. 443-459

Zuzovsky, R. (1987) *The elementary school in Israel and science achievement*. Ph.D. thesis. (Jerusalem, Hebrew University [Hebrew])

Zuzovsky, R. (1993) *Supporting science teaching in elementary schools within the policy of extended learning day* (Implementation Study, Center for Curricular Research and Development, School of Education, Tel Aviv University [Hebrew])

Zuzovsky, R. (1997) *Science in the elementary schools in Israel*. Findings from the 3rd International Mathematics and Science Study and a complementary study in Israel. (Center for Science and Technology Education, Tel Aviv University [Hebrew])

Zuzovsky, R., and Aitkin, M. (1993) *Final report: An indicator system for monitoring school effectiveness in science teaching in Israeli elementary schools*. Israel Foundation Trustees Grant 90 and the Chief Scientist's Office. (Jerusalem: Ministry of Education and Culture [Hebrew])

SUBJECT INDEX

NAME INDEX

Science & Technology Education Library

Series editor: Ken Tobin, *University of Pennsylvania, Philadelphia, USA*

Publications

1. W.-M. Roth: *Authentic School Science.* Knowing and Learning in Open-Inquiry Science Laboratories. 1995   ISBN 0-7923-3088-9; Pb: 0-7923-3307-1
2. L.H. Parker, L.J. Rennie and B.J. Fraser (eds.): *Gender, Science and Mathematics.* Shortening the Shadow. 1996   ISBN 0-7923-3535-X; Pb: 0-7923-3582-1
3. W.-M. Roth: *Designing Communities.* 1997
          ISBN 0-7923-4703-X; Pb: 0-7923-4704-8
4. W.W. Cobern (ed.): *Socio-Cultural Perspectives on Science Education.* An International Dialogue. 1998   ISBN 0-7923-4987-3; Pb: 0-7923-4988-1
5. W.F. McComas (ed.): *The Nature of Science in Science Education.* Rationales and Strategies. 1998   ISBN 0-7923-5080-4
6. J. Gess-Newsome and N.C. Lederman (eds.): *Examining Pedagogical Content Knowledge.* The Construct and its Implications for Science Education. 1999
          ISBN 0-7923-5903-8
7. J. Wallace and W. Louden: *Teacher's Learning.* Stories of Science Education. 2000
          ISBN 0-7923-6259-4; Pb: 0-7923-6260-8
8. D. Shorrocks-Taylor and E.W. Jenkins (eds.): *Learning from Others.* International Comparisons in Education. 2000   ISBN 0-7923-6343-4
9. W.W. Cobern: *Everyday Thoughts about Nature.* A Worldview Investigation of Important Concepts Students Use to Make Sense of Nature with Specific Attention to Science. 2000   ISBN 0-7923-6344-2; Pb: 0-7923-6345-0
10. S.K. Abell (ed.): *Science Teacher Education.* An International Perspective. 2000
          ISBN 0-7923-6455-4

KLUWER ACADEMIC PUBLISHERS – DORDRECHT / BOSTON / LONDON